Legal Personhood and the Rights of Nature

This book develops a novel theory of legal personhood, providing a theoretical grounding for the Rights of Nature movement.

Maintaining that natural entities such as glaciers and ecosystems are deserving of rights of their own, the Rights of Nature movement has been credited with the potential to bolster environmental protections and help halt the ongoing destruction of our planet. This book fills a gap in the literature by explicitly examining whether there are any necessary or sufficient conditions for legal personhood, and if there are any constraints on who or what can be a legal person. The book ultimately argues that there are no such constraints, which removes any need for legal persons to resemble human beings. By rejecting the connection between legal persons and humans, it becomes possible to finally move past the anthropocentrism that has so far colored nearly all aspects of international environmental law and which has held back the possibility of giving nature rights of its own.

The book is intended for those interested in the legal and philosophical foundation of rights in general and how they can be used to give nature rights and personhood of its own in particular.

Eric M. Rubenstein, University of Akureyri, Iceland.

Legal Personhood and the Rights of Nature

Beyond Anthropocentrism

Eric M. Rubenstein

Routledge
Taylor & Francis Group
a GlassHouse Book

First published 2027
by Routledge
4 Park Square, Milton Park, Abingdon, Oxon OX14 4RN

and by Routledge
605 Third Avenue, New York, NY 10158

Routledge is an imprint of the Taylor & Francis Group, an informa business

A GlassHouse Book

British Library Cataloguing-in-Publication Data
A catalogue record for this book is available from the British Library

Library of Congress Cataloging-in-Publication Data
Names: Rubenstein, Eric M. author
Title: Legal personhood and the rights of nature : beyond anthropocentrism / Eric Rubenstein.
Description: 1. | New York : Routledge, 2026. | Includes bibliographical references and index.
Identifiers: LCCN 2026025766 (print) | LCCN 2026025767 (ebook) | ISBN 9781041235828 hardback | ISBN 9781041241157 paperback | ISBN 9781003740186 ebook
Subjects: LCSH: Juristic persons | Environmental law | Jurisprudence | Nature—Effect of human beings on | Environmental ethics
Classification: LCC K237 .R83 2026 (print) | LCC K237 (ebook)
LC record available at https://lccn.loc.gov/2026025766
LC ebook record available at https://lccn.loc.gov/2026025767

ISBN: 978-1-041-23582-8 (hbk)
ISBN: 978-1-041-24115-7 (pbk)
ISBN: 978-1-003-74018-6 (ebk)

DOI: 10.4324/9781003740186

Typeset in Sabon
by Apex CoVantage, LLC

Contents

About the Author

Eric M. Rubenstein received a PhD in philosophy from the *University of North Carolina*, Chapel Hill, 1996. He taught philosophy at Colgate University from 1996 to 2000 and then at the *Indiana University of Pennsylvania* (IUP) from 2000 to 2023. His research has focused on metaphysics and the history of philosophy, with numerous articles and two books. He has taught a wide range of classes over the years, including 20th-century analytic philosophy, ethics, ancient philosophy, logic, and metaphysics, among others.

In 2023, he moved to Iceland and completed an MA in Polar Law at the *University of Akureyri* in 2025. His interests in Polar Law are varied, including efforts to preserve and protect Antarctica through the Antarctica Treaty System and related legal regimes; the challenges to, and viability of, UNCLOS as climate change turns once-ice-covered regions into open seas; the interaction and interplay between Western science and Indigenous knowledge; and finally, the importance of recognizing and furthering the self-determination of Indigenous Peoples in the Arctic.

Recent publications include a coauthored paper on rights of nature in Antarctica, one on the preservation of aesthetic values in Antarctica, and a paper on how feminist epistemology provides a helpful framework for understanding traditional Indigenous knowledge.

He lives in Akureyri, along with his sheep, Marvin, and far too many electric guitars.

Acknowledgments

I wish to thank various people from Taylor & Francis, who have provided encouragement and understanding through the various stages of completing this book. Thanks here to Colin Perrin for his advice and help in getting the project off the ground and for assistance at various points along the way. Laura Slater in particular deserves special thanks for her support and willingness to answer my seemingly endless string of questions and my frantic email ramblings. I've received various forms of feedback and constructive criticism, which have (I hope) greatly improved the quality of the final result. Mary MacLeod read an early draft of this book and provided the kind of insightful and pointed feedback I've come to know from her after over 30 years of friendship and as philosophy coconspirator. Special thanks as well for all of her support over these many long years.

I would like to acknowledge and thank numerous authors, journals, and publishers for permission to make use of previously published material. In particular, with permission from University of Chicago Press, I have quoted from Nurit Bird-David's "Animism Revisited: Personhood, Environment, and Relational Epistemology," volume 40 of *Current Anthropology,* 1999; with permission conveyed through Copyright Clearance Center, Inc. The *Global Journal of Animal Law* has granted permission to quote from Jonas-Sébastien Beaudry's "From Autonomy to Habeas Corpus: Animal Rights Activists Take the Parameters of Legal Personhood to Court," volume 1 of *Global Journal of Animal Law*, 2016. I have quoted from Patrik Baard, "Rights of Nature Through a Legal Expressivist Lens: Legal Recognition of Non-Anthropocentric Values," *Ethical Theory and Moral Practice,* volume 28, through a Creative Commons License. With permission from the Yale Law Journal Company, Inc., I have quoted from John Dewey, "The Historic Background of Corporate Legal Personality," *Yale Law Journal*, volume 35, number 6, 1926, with permission conveyed through Copyright Clearance Center, Inc. The Georgetown University Law Center has given permission to quote from Susan Emmenegger and Axel Tschentscher, "Taking Rights Seriously: The Long Way to Biocentrism in Environmental Law" in *Georgetown International Environmental Law Review*, © 1994, volume 6(3). With permission from the publisher Taylor & Francis Ltd., I have used material from Anna Grear's "Law's Entities: Complexity, Plasticity and Justice," *Jurisprudence* 4(1), 2013. Taylor & Francis Ltd. has provided permission to quote from Rachel Killean and Damien Short, "A Critical Defence of the Crime of Ecocide," *Environmental Politics*, 2025. Making use of Creative Commons Attribution-ShareAlike BY-SA 3.0 IGO, I have quoted extensively from UNESCO and their entry on Laponia as a World Heritage Site. Hart Publishing, an imprint of Bloomsbury Publishing Plc., has granted permission to use quoted material from Ngaire Naffine's *Law's Meaning of Life: Philosophy, Religion, Darwin, and the Legal Person,* 2009. The publisher Oxford Publishing Limited has provided permission to quote from Jonathan Raz, *The Morality of Freedom*, 1986. Quoted material is reproduced with permission of the licensor through PLSclear. With permission from Cambridge University Press, I have quoted from Mihnea Tănăsescu, "Rights of Nature, Legal Personality and Indigenous Philosophies," *Transnational Environmental Law*, volume 9(3), pp. 450–1, 2020. Courtesy of Taylor & Francis Ltd. I have permission to reprint material from Tim Ingold, "The Temporality of the Landscape," *World Archaeology*, volume 25(2), 1993, with permission conveyed through Copyright Clearance Center, Inc. Finally, Anthony Weston has kindly permitted use of material from his "Beyond Intrinsic Value: Pragmatism in Environmental Ethics," *Environmental Ethics,* volume 7(4), 1985.

My time in Iceland has been graced with the friendship and support of many people. I am especially thankful to Klara Káradóttir, Laura Buiret, Anna Lauenburger, Mitchell Sallis, Niall Janssen, and Hrafnhildur Fönn Ingjaldsdóttir. Special thanks as well to various friends from around the world who have supported my rather radical changes in life, particularly my decision to pack up my cozy life in the US and move to a (beautiful) rock in the middle of the North Atlantic. Diana Savage Thompson and Nicole Goulet deserve special mention here.

Finally, I wish to thank my family for their ongoing love and support through the vortex of vicissitudes, that is life. Though they perhaps have never quite understood why I've been drawn to the academic side of life, nor why certain issues and debates fascinate me, their support has been unwavering.

Preface

This book began life as a thesis written to complete a degree in Polar Law at the University of Akureyri, Iceland. The thesis was directed by Romain Chuffart, who at the time was the Nansen Professor in Arctic Studies. My introduction to Polar Law, and international law more generally, was through Romain, and he has arguably had the greatest influence on my thinking about these matters. I am most grateful for his tutelage and friendship. I was fortunate as well to have been guided in my studies by other faculty here in Akureyri, particularly Rachael Lorna Johnstone and Antje Neumann.

My interest in debates about legal personhood was sparked by Visa Kurki's excellent book, *A Theory of Legal Personhood*. From there, I encountered the writings of Ngaire Naffine and Anna Grear. Those three have provided the foundation for my thinking about persons and personhood and exemplify what I've always believed about succeeding in academic pursuits: often, you need a good bit of luck to find projects that speak to you. Encountering their writings definitely counts as good luck, though it should go without saying that they bear no responsibility for the consequences of these encounters. Visa Kurki has also been kind enough to provide helpful feedback and encouragement at early stages of this project.

My first love as an academic was philosophy, and over the years, I found myself repeatedly drawn in by the most abstract of ideas and debates. That interest in abstract, theoretical debates clearly has stayed with me, even though the topics I'm drawn to have changed over the years. What is perhaps new for me has been an interest not only in the theoretical but also the practical consequences and implications that flow from the theoretical. I hope that these dual interests become clear as the reader moves through the book—beginning with the most abstract of debates about personhood, rights, and nature and eventually moving to how some of those debates play out in finding new ways to address one of the most pressing issues of our time—the ongoing destruction of our planet.

The journey from working in philosophy, which I taught for over 20 years as Associate Professor at the Indiana University of Pennsylvania, to studying international law has been a challenge. An exciting one, for sure, but definitely daunting. So, too, has been simply being a student again after spending nearly 30 years on the teaching side of the classroom.

The present work focuses on the prospect of there being rights of nature and of nature being a legal person. Particularly challenging has been the need to put to the side intuitions about the grounds for *moral* rights of nature, focusing instead on the prospects of *legal* rights and *legal* personhood. Old habits die hard. Thus, I ask for the reader's patience in those instances when I find the language of philosophy a more familiar way of making the various points I chase here. At various junctures, I wrestle with the question of identity over time for things like ecosystems and forests. Only recently has it dawned on me that similar questions arise over my own identity over time as I've moved to a foreign country, became a student again, pursued another degree, changed disciplines, and formed new friendships. For better or worse, the present work is a result of the confluence of all those forces.

Introduction

In the 1972 United States Supreme Court case, *Sierra Club v. Morton*, the Sierra Club sued federal officials to stop the construction of a large ski resort in Mineral King Valley, California. The Sierra Club argued, unsuccessfully, that the resort would harm the surrounding wilderness. Though the Supreme Court ultimately ruled that the Sierra Club lacked standing to bring the case, two crucial points for our purposes emerged. The first, from Justice Blackmun, was a nod to the creative possibilities of law, possibilities we must explore and develop if law is to evolve in a way that allows it to meet emerging socio-economic-political challenges. In many ways, this gesture towards the possibilities of law provides the overarching theme of the pages and arguments that follow throughout this book. Blackmun wrote, in dissent:

> [T]his is not ordinary, run-of-the-mill litigation. The case poses—if only we choose to acknowledge and reach them—significant aspects of a wide, growing, and disturbing problem, that is, the Nation's and the world's deteriorating environment with its resulting ecological disturbances. Must our law be so rigid and our procedural concepts so inflexible that we render ourselves helpless when the existing methods and the traditional concepts do not quite fit and do not prove to be entirely adequate for new issues?[1]

The second crucial point, this time in a dissent from Justice Douglas, recognizes the possibility of giving nature itself its own rights and legal personhood, echoing the long-standing practice of giving rights to inanimate objects, as needed for resolution of any of a number of legal issues.

> Inanimate objects are sometimes parties in litigation. A ship has a legal personality, a fiction found useful for maritime purposes. The corporation sole—a creature of ecclesiastical law—is an acceptable adversary, and

1 *Sierra Club v. Morton* (1972) 405 U.S. 727, 755–756 per Blackmun J. (dissenting).

DOI: 10.4324/9781003740186-1

> large fortunes ride on its cases. The ordinary corporation is a "person" for purposes of the adjudicatory processes, whether it represents proprietary, spiritual, aesthetic, or charitable causes. So it should be as respects valleys, alpine meadows, rivers, lakes, estuaries, beaches, ridges, groves of trees, swampland, or even air that feels the destructive pressures of modern technology and modern life. The river, for example, is the living symbol of all the life it sustains or nourishes—fish, aquatic insects, water ouzels, otter, fisher, deer, elk, bear, and all other animals, including man, who are dependent on it or who enjoy it for its sight, its sound, or its life. The river as plaintiff speaks for the ecological unit of life that is part of it. Those people who have a meaningful relation to that body of water—whether it be a fisherman, a canoeist, a zoologist, or a logger—must be able to speak for the values which the river represents, and which are threatened with destruction.[2]

These possibilities, from Blackmun and Douglas, as well as the theories and controversies which lay behind them, will be discussed at length in the pages that follow. By way of beginning, however, we should take further note of Blackmun's dissent. Namely, as I read Blackmun, law is not static, nor should it be. If nothing else, law is a tool for mediating social problems, and to meet the challenges a society faces, law must be willing to make use of its creative possibilities. Law without change and evolution is destined to be mere dead letter, of no use to anyone.

With that in mind, this book seeks to draw on law's creative possibilities to meet one of the most urgent problems of our time, the very problem Douglas and Blackmun were pointing to over 50 years ago—the ongoing degradation and destruction of our environment. In what follows, I focus in particular on the potential for grounding the standing and potential rights for natural objects in a conception of legal personhood that escapes the traditional limits that have often been placed on the concept—namely, that legal persons must resemble the human person as it is variously conceived by biology, morality, and even metaphysics. In its place, I seek to broaden and extend the concept of legal personhood, in a way that buttresses the potential for how nature itself can be protected in law.

Douglas also explicitly mentions Christopher Stone's famous arguments for giving natural objects such as trees legal standing.

> The critical question of "standing" would be simplified and also put neatly in focus if we fashioned a federal rule that allowed environmental issues to be litigated before federal agencies or federal courts in the name of the inanimate object about to be despoiled, defaced, or invaded by roads and

2 *Sierra Club v. Morton* (1972) 405 U.S. 727, 755–756 per Douglas, J. 742–744 (dissenting).

> bulldozers, and where injury is the subject of public outrage. Contemporary public concern for protecting nature's ecological equilibrium should lead to the conferral of standing upon environmental objects to sue for their own preservation. *See* Stone, Should Trees Have Standing?—Toward Legal Rights for Natural Objects, 45 S.Cal.L.Rev. 450 (1972). This suit would therefore be more properly labeled as *Mineral King v. Morton.*[3]

Stone's famous essay, as well as Blackmun's and Douglas's dissents, are framed in terms of legal standing. I will focus on the notion of legal rights, not standing. This is for three reasons. First, a focus on the potential for natural objects to have legal standing is a product of English common law, as it is practiced in such places as the United States and Canada but arguably not the central legal concept in other jurisdictions.[4] Second, and more importantly for our purposes, as the focus of this book is not on comparative issues across domestic legal systems, is that speaking of potential *rights* for nature is much in vogue today, with hundreds of initiatives around the globe aimed at giving rights to natural objects such as rivers, forests, and ecosystems. Thus, to cast a wider net, and to engage with issue of environmental protection framed in terms of rights, that will be my focus. Indeed, many today view the multitude of initiatives to give nature its own rights as common and widespread enough to speak of "the Rights of Nature movement" (RoN). If for no other reason, the shared goal of giving rights and personhood to nature across the many initiatives is enough for practical purposes to call it a movement, even if we don't explore a precise definition of what counts as a "movement" per se.

Third, for someone (or some *thing*) to have legal standing requires, among other things, showing that they have suffered a harm to their interests—or that such a harm is imminent. As we will see in Chapter 4, one of the central challenges to giving nature rights is to show that it is indeed the kind of thing that can be harmed—in the legal sense. Thus, to speak of possible standing is perhaps to already presume nature can be harmed, which is something that needs to be defended, not assumed, particularly as so many critics (and even defenders) of RoN have pointed out it is by no means obvious that nature is the kind of thing that can be harmed, given that it is not sentient.

Three Grades of Legal Visibility

To coin a phrase, we might say the importance of legal personhood is that it is one of several ways in which something can become *legally visible*. To be

3 *Sierra Club v. Morton* (1972) per Douglas, J. 741.

4 Gwendolyn McKee, 'Standing on a Spectrum: Third Party Standing in the United States, Canada, and Australia' (2011) 16(1) *Barry Law Review* Article 4. <https://lawpublications.barry.edu/barrylrev/vol16/iss1/4>

visible in the legal sense is to be recognized by law as deserving of a certain kind of respect. To be legally visible mandates law take a being's existence seriously in a way that the legally invisible are not. For while a piece of property may be significant to the parties in a dispute over ownership, that property is merely what the conflicting rights (or duties) are about, and in that way, it is *not* legally visible in the way that the parties to the dispute are. Put differently, that portion of land is merely a legal *object*, not a *subject*.

Now being legally visible can occur in at least three ways. On what I'll call the *first degree* of visibility, something can be worthy of respect in the sense that there have been laws or judicial rulings that signify that though we are dealing with only a legal object, that object shall be protected from possible harm. Here, I have in mind the large array of efforts, domestic and international, that seek to protect the environment. These may take the form of conservation requirements, intended to protect the biodiversity of the land and sea, perhaps in the name of preserving them for future generations, or in the form of prohibiting development or the exploitation of natural resources found on those lands. What things at this first level or degree of visibility have in common, again, is they are special enough to warrant legal protections in a way that sets them apart from other regions of the natural world.

Put differently, on this first degree, nature may be protected because it is a special kind of object. It is still merely an object, not subject, and thus won't have any rights of its own. Still, it is deemed important enough that it shows up in law as deserving of special treatment. This can be seen, for example, at the domestic or national level, where States set aside certain natural areas as parks. At the international level, a prime example can be found in the 1991 Environmental Protocol to the Antarctica Treaty, which, in speaking of the special values of Antarctica, spells out the various ways Antarctica and its values are to be protected, including the specification of types of environmental impact assessments that are required before any development occurs, such as construction of research centers.[5]

Next, we move to what I consider the *second degree* of visibility, where nature is given specific rights, such as the right to exist, to flourish, and to be restored.[6] By being directly afforded particular rights, an entity is encoded into law such that it now has a different kind of visibility than merely being a specially protected object. For in being given rights (and/or duties), we cross an important threshold whereby something becomes a legal *subject* and no longer merely a legal object. Legal objects don't have rights, after all, and in

5 The Antarctic Treaty, 402 U.N.T.S. 71, *entered into force* June 23, 1961; 1991 Protocol on Environmental Protection to the Antarctic Treaty 30 ILM 1455 (1991), otherwise known as "Madrid Protocol."

6 As an example, as we'll see, Ecuador's 2008 Constitution provides specific rights to nature to *exist* and *flourish*.

giving, say, particular regions of nature, or nature as a whole, rights of its own, we are indicating another level in which law has specified or picked out something as importantly different, something that is to be respected and protected in the way that legal rights provide protection.[7]

Finally, and most importantly for our project, we have a *third degree* of visibility where something is explicitly given legal personhood. Here, when *explicitly* recognized as a legal person we have, I suggest, the strongest, most powerful way of being legally visible. As I see it, with the second degree of visibility that comes in directly giving to something, we are *implicitly* giving in personhood as well. But, as I say, that is granting it only implicitly. The thought here is that because non-persons don't have rights, to give something rights is to at least implicitly recognize it as a legal subject or legal person.[8]

When rights are bestowed in the manner of the second degree of visibility, they tend to be narrowly defined and focused. In contrast, the bestowal of personhood and visibility in the *third* degree, where personhood is *explicitly* given, provides for a broader range of rights than we find in the second degree. In addition to the breadth, we also can speak of the *depth* of that visibility. Deciding something is a legal person is a way of saying that entity is in the legal domain in a unique way that makes it of particular importance. It is, in essence, to say it is one of *us*, the kind of being that matters in the law. And when that is done explicitly, as I say, I think we have found the strongest, most powerful way law can signify a being's importance. The term "person," after all, carries significant weight in morality and in law. Being a *person* is to stand apart as something uniquely important, both in its value and in how it is to be treated.

With all that in mind, I turn to the particular focus of this book, which is to examine the Rights of Nature (RoN) movement, which seeks to establish the legal rights of non-human nature, rights that nature should have in and of itself. RoN can be seen, that is, as attempting to give nature legal visibility. As I've noted, that can come in different versions, and I'm particularly concerned about the latter two I've described, namely, giving pride of place to particular rights, which implicitly implicates something as a legal person, or in the other direction, of explicitly granting something personhood from which a wide range of rights (and duties) may flow.

As we proceed, we'll see that each of these routes have been taken for nature. Some RoN initiatives have focused on particular rights for nature, while others have sought to directly accord nature legal personhood. And though at times I'll be speaking of nature's *rights*, the ultimate aim is to explore the extent to which there are or should be constraints on granting nature the

7 The various ways rights are understood as giving protection will be explored later when we examine more closely the function and purpose of rights themselves.

8 I am using "legal subject" and "legal person" interchangeably, whereas others may not.

status of being a *legal person*, for as I've suggested, this is the deeper kind of legal visibility. Can nature be a legal person, in other words? Does it have "what it takes" to be worth of that highest degree of legal visibility?

Realism and Legalism

As said, RoN seeks to establish the legal rights of non-human nature, rights that nature has in and of itself. While various initiatives worldwide differ on the types of rights, legal standing, and methods of representation of nature, they have in common the idea that natural entities or beings deserve legal protection irrespective of human interests.[9]

Importantly, there are various motivations in seeking to provide nature with its own legal rights. RoN can be considered, for instance, as a political struggle playing out among different groups in their particular legal and political contexts.[10] RoN initiatives may also be seen as furthering recognition of Indigenous Peoples' worldviews and ultimately to assist in their pursuit of self-determination.[11] Alternatively, RoN might simply be a pragmatic measure, one that seeks to increase environmental protections, even if the ultimate goal is to establish those protections primarily for the interests of humans.

Whatever the motivation, given the deep connection between something having rights and being a legal person, it becomes crucial to ask whether nature is indeed the kind of thing that can be the subject of rights, that can be the kind of thing that is a legal person. RoN answers in the affirmative. But that view also has many detractors. In hopes of adding to the multitude of arguments already in play on behalf of RoN, I draw on an even longer-standing debate within legal theory, the debate over the nature of legal personhood itself and what kind of constraints in general, if any, there are for being a legal person.

Here, I draw on terminology from Ngaire Naffine, who offers a helpful way of framing those debates about the nature of personhood. On her view, which I discuss at length in subsequent chapters, we should distinguish, on one side, those who view personhood as necessarily tied to one or more features deemed essential to the nature of human beings. These *Realist* views are to be contrasted with what she dubs the *Legalist* strategy, which argues that legal personhood can be severed from any metaphysical constraints, which

9 For an overview of initiatives worldwide, see 'Eco Jurisprudence Monitor,' a website that tracks various kinds of regional, domestic, and international efforts, on a range of topics from giving rights to grantings of personhood. <http://www.ecojurisprudence.org>

10 Mihnea Tănăsescu, 'The Rights of Nature as Politics' in DP Corrigan and M Oksanen (eds), *Rights of Nature: A Re-Examination* (Routledge 2021) 69.

11 Erin O'Donnell, Anne Poelina, Alessandro Pelizzon and Cristy Clark, 'Stop Burying the Lede: The Essential Role of Indigenous Law(s) in Creating Rights of Nature' (2020) 9 *Transnational Environmental Law* 403 <http://doi.org/10.1017/S2047102520000242>

in turn makes it possible for anything in principle to be a legal person. Legalism, in other words, can make something a legal person even if that entity has no resemblance to features found in human beings, what the law typically refers to as a "natural persons." On the Legalist view, "person," in the legal sense, has no necessary connections with persons in the physical, moral, or metaphysical sense.

Prima facie, the Legalist strategy seems promising for RoN, in that its open texture seemingly allows natural entities the potential for legal personhood. On that view, there would be no need to argue that nature has features relevantly similar to those found in physical human beings. The threshold, in other words, for deeming something a legal person is much lower on the Legalist view. If anything in theory can be a legal person (and in defense of that view, we often find reference to the range of exotic entities that have been granted personhood over the centuries), then it follows that nature could be one of things granted personhood.

There are, however, important objections to pursuing nature's personhood via the Legalist strategy. One in particular, stressed by Naffine and others, is that *how* the law actually operates shows its fidelity to the Realist position. Attempts to justify the use of "person" in the wholly open manner, Legalism suggests, say, by arguing that legal language can be isolated from the non-legal such that terms as "person" have a meaning wholly determined by its use in *law*, fail to reckon with the fact that the law often uses "person" in anything but an unspecified manner. Very often, in fact, critical debates and legal rulings rely on substantive conceptions of "person," focusing on various salient traits of human beings.[12] No account of the actual use of *person* in law, it seems, can vindicate the idea that jurists and RoN advocates are free to simply use "person" in whatever way they wish.

The challenge for the Realist who is sympathetic to RoN, on the other hand, is that attempting to argue that nature warrants standing as a legal person requires one to draw parallels between human beings ("persons" in the non-legal sense) and nature. To my mind, as I will argue, this often requires an unacceptable level of anthropomorphism. Relatedly, attempts to argue for nature's deserving to be a subject of law, as opposed to merely an object, often rely on moral arguments aiming to demonstrate the intrinsic value of nature. Here, it is commonly held that *all and only* beings with moral worth or intrinsic value are suitable candidates for personhood.

12 Ngaire Naffine, *Law's Meaning of Life: Philosophy, Religion, Darwin, and the Legal Person* (Hart 2009) 79. Having a human soul is another example of a feature deemed necessary for personhood, a position defended by Finnis, among others, in the Christian tradition of natural law. See, for example, John Finnis, 'Natural Law Theories' *Stanford Encyclopedia of Philosophy* <https://plato.stanford.edu/entries/natural-law-theories/>

The problem for this strategy, which I will argue is another variant of Realism, is that it is quite difficult to provide a coherent conception of *intrinsic value* that would include sentient animals along with the non-sentient and the a-biotic realms of nature. Arguing that animals, for instance, have intrinsic value or moral worth is challenging enough, though at least there we have a possible connection with features of humans we do think valuable, be it via the intelligence, or less restrictively, the sentience of animals. But even that will only get us so far in the biologic realm, leaving out many living things. Finding a way to argue that the *a-biotic realm* also has intrinsic value seems a bridge too far. Thus, the entire edifice of the Realists' justification for nature's legal personhood hangs on rather precarious premises.

With the earlier discussion as background, this book has two overriding goals. The first can be understood as defending the hypothetical or conditional claim that *were* nature to be granted rights and personhood, then all things considered, its rights and personhood should be grounded in Legalist rather than Realist foundations. I will argue, again, that a Legalist approach to nature's personhood is theoretically preferable to the various Realist views that have been developed by advocates of RoN. On balance, I believe there to be satisfactory replies to objections to the Legalist view, ones that are stronger than possible replies on behalf of the Realist. The second goal is to establish the antecedent, as it were, of that conditional claim. Nature should, in other words, be given personhood and rights of its own. The heavy lifting in establishing that nature should be given rights and personhood is done via the notion of *interests*, which, in Chapter 4, I argue can be understood in a way that, in keeping with Legalist scruples, does not require a linkage with either sentience or intrinsic value, as various defenders of interest theories of rights have held.

In developing and defending what I'm calling a "novel theory of personhood," I will distinguish two different versions of Legalism, which can be understood as two ways in which we might view the concept of a legal person as independent and separate from various conception of human beings. (The contrast, again, being with "Realist" positions, which posit there is and must be a tight connection between the two.)

On what I'll call a "Strong Legalist" view, anything can be a legal person. This seems to be the view advanced by people like Dewey, Klabbers, and Kelsen, whom I discuss in Chapter 1. On the other end of the spectrum, we can speak of a "Weak Legalism," which claims that while it is indeed crucial to separate legal personhood from non-legal conceptions of persons, there are still limits on what can be a legal person. That is the view I defend.

Put differently, while Strong Legalism claims there are no *a priori* constraints on what can be a legal person, my Weak Legalism holds there are, in fact, *some* constraints on being a legal person, and those constrains make it false that anything can be a legal person. The constraint that I hold out as salient is that legal persons have to be the kinds of things that can be made better or worse off, which, as I've said, amounts to the claim that those things have interests. In

Chapter 4, I argue that similar to corporations, natural entities can be made better or worse off. Crucially, though, like corporations, nature has interests and can thus be made better or worse off without making sentience be part of picture.

Intrinsic Value and Anthropocentrism

As mentioned, RoN at times seeks grounding in the idea that nature has *intrinsic* value, a value akin to that traditionally thought to apply only to human-centered traits such as rationality or happiness. On this understanding, nature should have rights of its own, as it is the kind of thing that deserves or warrants that standing in virtue of its standing as something more than instrumentally good for human ends. In general, the linking of intrinsic value and rights is prima facie plausible, as we do tend to hold that things with intrinsic value deserve legal protections in the strongest of senses.

Recognition that nature has intrinsic value and thus should be protected in law for its own sake can be seen as part of the larger effort to move beyond the anthropocentrism that historically has framed environmental law and policy. Emmenegger and Tschentscher connect these various strands as follows, speaking not of nature in general but of the idea that *biodiversity* might be of intrinsic value.

> The assertion of nature's intrinsic value made its entry into the law-making process of multilateral environmental instruments only recently. By proclaiming that nature has a value which is independent of human interests, these multilateral instruments use a very different kind of argument and thereby express a paradigm shift in environmental law. The conceptual difference between recognizing non-anthropocentric value and evaluating all other kinds of anthropocentric values is best expressed in the following introductory sentence of the Biodiversity Convention: "Conscious of the intrinsic value of biological diversity and of the ecological, genetic, social, economic, educational, cultural, recreational and aesthetic values of biological diversity and its components, . . . [we have] agreed as follows: . . .". This passage not only draws a clear line between intrinsic value of nature on the one hand, and all kinds of anthropocentric values on the other, but also puts non-anthropocentrism first, thereby emphasizing its importance as a new approach. For these reasons, we take intrinsic value recognition as a distinctive third stage for our overall thesis about a step by step development of international environmental instruments.[13]

13 S Emmenegger and A Tschentscher, 'Taking Nature's Right's Seriously: The Long Way to Biocentrism in Environmental Law' (1994) VI(3) *Georgetown International Environmental Law Review* 545, 568. Reprinted with permission of the publisher, Georgetown International Environmental Law Review © 1994.

On their view, recognizing the intrinsic, non-instrumental value of nature marks a crucial step in moving beyond the human-centered orientation in environmental law and policy. And to the extent that RoN is grounded in the idea that nature has its own value, independently of being a resource for human, RoN itself can also be seen as taking an important step away from anthropocentrism.

Earlier, I mentioned my skepticism about grounding nature's legal rights in moral rights and in the notion of intrinsic value. That will be developed in Chapter 3. But that abandonment of value talk might seem to create problems for the anti-anthropocentric view that I think rightly characterizes RoN. The very idea of connecting RoN with the developing notion of moving beyond non-anthropocentrism seems to be that nature has non-instrumental value and that that value should be respected, if not put at the center of environmental law and policy. But draining talk of intrinsic value out of non-anthropocentrism seems to leave little left to the latter. And that seems like a mistake. For if we eschew talk of intrinsic value, and nature having intrinsic value, what exactly is it that we are claiming should be reclassified not as anthropocentric but as non-anthropocentric?

By way of reply, and to show how it is true that RoN indeed represents a move away from anthropocentrism in environmental law and policy, without relying on the notion of intrinsic value at all, I will suggest that nature's alleged intrinsic value is neither sufficient nor even necessary for it to have rights of its own. On the insufficiency, recall our three degrees of visibility. Were we to acknowledge that nature does have intrinsic value (which I won't), that is not sufficient to give it legal rights. Moral rights and evaluative notions about value are distinct from legal rights, though, of course, they may be related.

More importantly, showing that a specific region of nature, or even perhaps biodiversity itself, has intrinsic value might still leave us only at what I called the first degree of legal visibility—nature as an *object* of law, though one that has been designated as special in some way and in some way deserving of protection. Indeed, that looks be exactly what we find in various initiatives concerned to articulate the intrinsic value of nature, including the Biodiversity Convention quoted earlier.

For that Convention, even in recognizing the intrinsic value of nature does *not* call for biodiversity or regions with high levels of biodiversity to be given rights of their own. It simply calls for biodiversity to be protected, i.e. to have an initial or first degree of visibility, in my terminology.[14]

14 Again, there is not a call for biodiversity to have rights of its own, but instead, it is to be conserved as a resource. "The objectives of this Convention, to be pursued in accordance with its relevant provisions, are the conservation of biological diversity, the sustainable use of its components and the fair and equitable sharing of the benefits arising out of the utilization

We can find similar characterizations and connections in the Environmental Protocol to the Antarctica Treaty, which speaks, *inter alia*, of the intrinsic value of Antarctica and the requirement to protect it but notably does *not* speak of Antarctica as having rights of its own.

> The protection of the Antarctic environment and dependent and associated ecosystems and the intrinsic value of Antarctica, including its wilderness and aesthetic values and its value as an area for the conduct of scientific research, in particular research essential to understanding the global environment, shall be fundamental considerations in the planning and conduct of all activities in the Antarctic Treaty area.[15]

Now Emmenegger and Tschentscher are surely right that recognizing nature's intrinsic value would mark an important step towards a non-anthropocentric worldview. I want to suggest, however, that an even greater step towards supplanting humanity as measure of all that is important comes in our second and third degrees of visibility, where rights and personhood are given to nature. That would truly be a major paradigm shift, as it would recognize nature, in law, as protected on a par with how humans and human interests are protected.[16] By giving nature at least second, but even better third degree of visibility, we elevate it as something protected by law in the strongest way that the law allows—of personhood with rights. It would be non-anthropocentric in perhaps an even deeper way than a recognition of nature as a special kind of object, in that it extends rights to entities beyond humans and recognizes those entities as subjects of law, not mere objects.

Finally, on the Legalist view I am defending, nature has these legal rights and personhood not because it resembles humans or because it serves human ends. The non-human parts of nature don't need to be like humans in any way to be acceptable candidates for rights and personhood. *That* would be a genuine way of removing humanity from the center of things—of finding nature with rights of its own, regardless of whether it serves human interests and regardless of whether it looks anything like humans.

In short, our second, and more powerfully, our third degree of visibility, would be a truly powerful non-anthropocentric position. This could also be

of genetic resources, including by appropriate access to genetic resources and by appropriate transfer of relevant technologies, taking into account all rights over those resources and to technologies, and by appropriate funding." United Nations, *Convention on Biological Diversity* (1992) <https://www.cbd.int/convention/text/>. It was signed in Rio de Janeiro on June 5, 1992, and entered into force on December 29, 1993.

15 Protocol on Environmental Protection to the Antarctic Treaty, Article 3.

16 As I discuss in later chapters, clearly, this does not entail that the non-human parts of nature will have the same rights as the human portions.

put as the claim that nature's alleged intrinsic value is not even *necessary* for it to be deserving of rights and personhood. We will see that on my Weak Legalism, it is not intrinsic value that is key to having rights, as many have claimed, but rather, it is that nature has interests. Those interests, furthermore, can be defined in a way that does not implicate any human-like feature, such as sentience, nor does it require a thing with interests to have intrinsic value.

"Nature" versus "Nature"

A final few important introductory remarks are in order. In Chapter 4, I argue that when distributing rights to nature or encoding nature as a legal person, it is preferable to speak of particular regions of nature, such as particular ecosystems, instead of aiming to cover all of nature, considered in a broad overarching manner. That still invites the question of how I understand the nature of *nature*. This becomes particularly pressing given that in Chapter 5 and 6, I examine arguments which seek to undermine the commonly made distinction between nature and culture.

Providing a definition of "nature" is rather tricky, given that in one sense, everything that exists is natural. In that way, "nature" is all inclusive, as there is nothing that is beyond nature. Perhaps more important than a definition, however, is how one might understand the goals of the RoN project itself. Humans after all are a part of nature. We have physical bodies subject to the laws of physics, chemistry, and biology. We are, in other words, an aspect of nature that has decided to invoke a range of rights and duties on *one* part of nature, namely, ourselves. In what follows, I will often speak simply of "nature" and whether it is appropriate to give it rights and/or personhood. A fuller statement, which I don't use everywhere for ease of expression, is that we are confronted with the challenge and possibility of extending rights and personhood to the non-human parts of nature.

Law obviously protects fundamental human interests, be they rights to life and physical security, rights to speech and assembly, not to mention rights to form associations and to pursue economic interests. But can and should the various legal protections we provide for ourselves be extended beyond the human realm, by recognizing the personhood and/or rights of entities beyond humans? Whether we are talking about animals or rivers, plants or ecosystems, perhaps less important than a tidy definition of *nature* is asking who or what can count as something worthy of the highest of legal standings—i.e. of having rights and personhood. Again, can and should we extend some of the most powerful of legal tools, rights and personhood, beyond the domain and interests of humans? Can and should we widen the scope and domain of what should be given the legal protections we already give to ourselves?

When we look at the history of how rights have been distributed, and in particular who they have been distributed to, we find, in the phrase of Anna

Grear, deeply entrenched *patterns*. The historical pattern of taking the paradigm of a legal person to be a white, property-owning male has historically made it difficult to find room for considering women as full-fledged persons.[17] The racial element of legal personhood has likewise led to the exclusion of people of color. We might say that this entrenched way of construing personhood has led to the "othering" of those who don't fit the paradigm, where that has led to their alienation and exclusion from legal protections, namely, rights.

As we turn to consider the possibility of nature (beyond the human aspect) itself having rights, we might be aware of another entrenched pattern, perhaps one of the deepest of all, namely, the idea that would-be rights-bearers have to resemble the essential features of human beings. This requirement, as I've suggested, is the essence of the Realist approach. Should we wish, in contrast, to extend legal rights and personhood beyond the human realm, perhaps we need to find ways to see beyond and break that patten. This is the promise and potential of Legalism, I believe. For if there are no (or very few) *a priori* constraints on what can be a legal person, then being a right-bearing entity is not limited to those entities which look or sound like us. As I will argue in the following pages, it is the open possibilities of personhood provided by Legalism that is well suited to this task, unlocking the full power of the creative aspects of law to address the ongoing destruction of nature, which is one of the most pressing problems of our time. To close the circle, this is to find a way to *not* be trapped by our concepts, theories, history, and methods, as Justice Blackmun, quoted earlier, worried about.

Organization

This book proceeds from highly abstract discussions to increasingly concrete ones. Chapter 1 focuses on the theoretical debate over the proper way to understand legal personhood. I contrast the Legalist and Realist approaches with an eye to arguing that on balance, the Legalist approach better captures the way "person" is used in legal reasoning. To reach that goal, I make use of arguments found in contemporary legal theory, as well as in contemporary analytic philosophy (i.e. Anglo-American) philosophy of language. On my view, "person" in the legal sense simply denotes a bearer of rights and duties. The legal sense of "person," that is, can be severed from claims about "person" in the biological, philosophical, and ethical sense.

17 "It is of particular note, for example, that existing critical accounts suggest a highly patterned selectivity in relation to the characteristics of the empirical 'something' that law is willing to deploy or acknowledge as a legal subject—a patterned selectivity moreover reflecting identifiable characteristics and suppositions concerning the particularities of 'natural' human being of law." Anna Grear, 'Law's Entities: Complexity, Plasticity and Justice' (2013) 4(1) *Jurisprudence* 76, 89.

Chapter 2 turns from the descriptive account of "person" in the first chapter to examine the normative side, investigating not just how "person" can be understood but how it should be. In seeking to maintain that nature is the kind of being that is fully capable of being a legal person, I critique Realist assumptions that would grant personhood to nature. Here, I argue that Realist accounts founder, as they tend to rely on problematic anthropomorphic conceptions of nature. Importantly, it is not the Realist advocacy of RoN that is the problem. Rather, it is using Realism as the foundation, thereby opening the door arguments in later chapters that a Legalist foundation is superior.

Chapter 3 continues the critique of Realist accounts of personhood, particularly as they are wielded in support of RoN. The focus in this chapter is on attempts to secure legal rights via moral considerations, in particular that nature has intrinsic value. I explore more fully the concept of intrinsic value, as well as what I take to be its unsuitability as a premise in arguing that nature is deserving of rights of its own. I conclude that in light of the challenges for the Realist approach on behalf of RoN, an alternative is worth exploring. The task of developing such an account emerges as the goal for subsequent chapters. A crucial part of the project is to push back against those who think nature lacks what is needed to be rights-bearer. Raz, in particular, thinks nature fails on account of its not being something that has "ultimate value." I discuss at length the weakness of this notion and how it is not able to do the excluding Raz seeks.

Chapter 4 examines the problematic notion of *interests*, for many have argued against nature's being a legal person on the simple grounds that nature seemingly is not the kind of thing that can have interests. If it doesn't have interests, then it isn't the kind of thing that can be harmed and thus not the kind of thing it makes sense to give rights to. Here, I suggest that the alleged obstacle again turns on assumptions that nature's interest must resemble the interests of human beings, particularly the idea that interests require sentience. But those are just Realist assumptions, and freeing ourselves from that assumption, by examining the way in which non-sentient corporations have interests, and by exploring a broader notion of interest, can help secure the idea nature can indeed be harmed and is thus suitable for having rights of its own.

Arguing, though, that nature has interests inevitably raises the question of "which nature"? Should we, that is, speak of individual trees, forests, rivers, or ecosystems or something more broad and abstract, namely, the protection of nature itself? Examining this contested issue leads me to conclude that on balance, we are better off speaking of particular portions of nature, avoiding the temptation to speak simply of Nature, capital "N."

In an appendix to Chapter 4, I examine the implications of these reflections on personhood by seeing how they frame the current debate over whether AI systems can be legal persons. On balance, even with the low bar I set for being a legal person, I tentatively conclude that AI is not a suitable candidate for personhood.

Chapter 5 puts the theoretical conclusions of the first four chapters to work in a test case, where I examine how a Legalist account of nature's personhood might fare in an Arctic context. As there are relatively few pending efforts to give nature personhood in the Arctic, unlike in other parts of the world, I propose a hypothetical case. In particular, the Laponia World Heritage Site in northern Sweden is a legally protected area governed by various laws intended to protect the natural and cultural history of the region. As it presently stands, though, while there are provisions in place regarding nature, they are not ones that give rights or personhood to Laponia itself. Exploring what the present legal regimes are capable of, and where they may fall short, I suggest how giving Laponia rights of its own might strengthen the various protections that are already in place. Chapter 5 (and 6 after it) become exercises in trying to be more exacting about how exactly the legal landscape changes—for both the entity and its environment—as its legal visibility increases.

Chapter 6 continues this exploration, moving from an examination of how things might be different were Laponia given its own rights to examining the possibilities that are opened in giving Laponia full legal personhood. That is, making use of the distinction in the ways something can be legally visible, I examine how a direct bestowing of personhood on Laponia could open the door for the introduction and adaptation of new rights, ones suited to the nature of Laponia itself. The possibilities of these developments are argued to be better facilitated by a Legalist understanding of Laponia's personhood.

Chapter 7 examines the implications of my Weak Legalism by exploring objections to it, particularly those which claim that in setting the bar so low for personhood—in simply requiring nature to have interests—I am forced to accept as legal persons other entities which prima facie have no business being considered persons.

Chapter 8 is the conclusion of these investigations. Ultimately, I argue that a Legalist understanding of personhood fares better than a Realist construal for advocates of RoN. That is one of the two larger goals of this book—to show that a Legalist account of personhood is better suited to the goals of RoN than a Realist one and that nature does have what it takes to be a legal person and rights-bearing entity, which is the conclusion of Chapter 4. I conclude with the semi-optimistic thought that RoN, on Legalist foundations, has a potential for greater effect than other efforts.

References

The Antarctic Treaty, 402 U.N.T.S. 71, *Entered into Force*, 23 June 1961.

Eco Jurisprudence Monitor <http://www.ecojurisprudence.org>

Emmenegger S and Tschentscher A, 'Taking Nature's Right's Seriously: The Long Way to Biocentrism in Environmental Law' (1994) VI(3) *Georgetown International Environmental Law Review* 545, 568.

Finnis J, 'Natural Law Theories' in *Stanford Encyclopedia of Philosophy* <https://plato.stanford.edu/entries/natural-law-theories/>

Grear A, 'Law's Entities: Complexity, Plasticity and Justice' (2013) 4(1) *Jurisprudence* 76, 89.

McKee G, 'Standing on a Spectrum: Third Party Standing in the United States, Canada, and Australia' (2011) 16(1) *Barry Law Review*, Article 4 <https://lawpublications.barry.edu/barrylrev/vol16/iss1/4>

Naffine N, *Law's Meaning of Life: Philosophy, Religion, Darwin, and the Legal Person* (Hart 2009).

O'Donnell E, Anne Poelina A, Pelizzon A and Clark C, 'Stop Burying the Lede: The Essential Role of Indigenous Law(s) in Creating Rights of Nature' (2020) 9 *Transnational Environmental Law* 403 <http://doi.org/10.1017/S2047102520000242>

Protocol on Environmental Protection to the Antarctic Treaty 30 ILM 1455 (1991), otherwise known as "Madrid Protocol".

Sierra Club v. Morton (1972) 405 U.S. 727, 755–756 per Blackmun J. (dissenting) <https://supreme.justia.com/cases/federal/us/405/727/#materials>

Sierra Club v. Morton (1972) 405 U.S. 727, 755–756 per Douglas, J. 742–744 (dissenting) <https://supreme.justia.com/cases/federal/us/405/727/#materials>

Tănăsescu M, 'The Rights of Nature as Politics' in DP Corrigan and M Oksanen (eds), *Rights of Nature: A Re-Examination* (Routledge 2021) 69.

United Nations, Convention on Biological Diversity (1992) <https://www.cbd.int/convention/text/>. It was signed in Rio de Janeiro on June 5, 1992, and entered into force on December 29, 1993.

Chapter 1

A Functional-Legalist Account of "Personhood"

A central aim of this book is to argue that a Legalist account of personhood provides a better theoretical foundation for RoN than its Realist foes.[1] Thus, it behooves us to be clear in how we carve up the conceptual space around the term "person" and its relatives, i.e. personhood, natural person, and legal person. As we will see, while the debate over the proper account of personhood for *nature* may be a new development, the debate over the nature of personhood itself is far from new. As Naffine put it, "The largest dispute, the biggest intellectual and moral battle, is between those who say that law does not and should not operate with a natural conception of the person and those who say that it does and should."[2]

Legalism Defined

As often noted, the term "person" derives from the Latin *persona*, meaning a mask. In that sense, every person has (at least) one *persona*. Their persona is the face that meets the world, the world where that persona has a proper role to fill. (Think again of actors in a play.) From there, its meaning seems to have shifted. Over the centuries, the sense of person as *persona* gradually came to designate something around which classical liberalism developed and flourished—the *person* as autonomous agent, one able to rationally reflect on its goals, curtail its desires when needed to achieve those goals, and to play the role of the basic entity from which societies are formed—formed from their free, rational, and self-interested motives.[3]

1 For simplicity's sake, I will drop the use of "metaphysical" when speaking of the Realist position.

2 N Naffine, *Law's Meaning of Life: Philosophy, Religion, Darwin, and the Legal Person* (Hart 2009) 20.

3 Anna Grear, 'Law's Entities: Complexity, Plasticity and Justice' (2013) 4(1) *Jurisprudence* 76, 83. Compare this passage from Keeton: "The word *person* is derived from the Latin *persona*. This term has a long and and interesting history. Originally it meant simply a mask. Later, it denotes the part played by a man in life, and still later, the man who plays it. In this sense

DOI: 10.4324/9781003740186-2

The idea of a *person* as a self-determining being, one with the capacity for reasoning and agency, is surely indicative of "person" being understood along Realist lines. I return to that conception later. I think it useful, however, to begin with *persona* in its original sense of artifice and mask. For that captures in a nutshell the Legalist understanding of *person*. For the Legalist, the artifice is that *persona*, *qua* mask, conceals as irrelevant anything underlying that mask. *Person*, for the Legalist, is not concerned with what lies under the mask, as it were. Instead, what matters is how that *persona* displays itself and what roles it plays, not in theater, of course, but in law and the role it plays therein.

I now quote at length several clear statements of the Legalist understanding of person, reaching back to early articulations from Thibaut and Kelsen, respectively.

"By Person is meant whatever in any respect is regarded as the subject of a right: by Thing, on the other hand, is denoted whatever is opposed to person."[4]

> The concept of the legal person—who, by definition, is the subject of legal duties and legal rights—answers the need of imagining a bearer of the rights and duties. Juristic thinking is not satisfied with the insight that a certain human action or omission forms the contents of a duty or a right. There must exist something that "has" the duty or the right. In this idea, a general trend of human thought is manifested. Empirically observable qualities, too, are interpreted as qualities of an object or a substance, and grammatically they are represented as predicates of a subject. This substance is not an additional entity. The grammatical subject denoting it is only a symbol of the fact that the qualities form a unity. The leaf is not a new entity in addition to all the qualities—green, smooth, round and so on—but only their comprehensive unity.[5]
>
> A legal person is the unity of a complex of legal obligations and legal rights. Since these obligations and rights are constituted by legal norms (more correctly: *are* these legal norms), the problem of "person" is in the last analysis the problem of the unity of a complex of norms. The question is: what constitutes this unity in the one case and in the other? . . . It is not

every human being has a *persona*. In later Roman law, however, the term acquires a still more specialised meaning. Becoming almost synonymous with *caput*. Thus a slave or an *impubes* being an imperfect *persona*. Last of all, the term comes to denote a being capable of sustaining legal rights and duties." George W Keeton, *The Elementary Principles of Jurisprudence* (Sir Issac Pittman 1949) 117.

4 Anton Thibaut, *An Introduction to the Study of Jurisprudence* (Nathaniel Lindley tr, T & JW Johnson 1885) §101.

5 Hans Kelsen, *General Theory of Law and State* (Harvard University Press 1945) 93.

> a natural reality but a legal construction, created by the science of law—an auxiliary concept in the presentation of legally relevant facts.[6]

More recent statements can be found in Klabbers, who summarized Von Savigny as holding that "the legal person is distinct from the natural person, and lacks any innate personality . . . has no will, no mind, and no ability to act, except to the extent that the law imputes such will and ability."[7]

Derham offers an important, often quoted perspective:

> The misconception is that there is a logical difference between a corporation as a legal person and a human being as a legal person. There is no such difference. In both cases the use of the term "legal person" is as necessary a part of a statement which if true or possibly true asserts, claims, decides, or assumes that a designated entity (whether a man, a group of men, a fund, an idol, or a purpose) is treated, under the rules of the legal system with reference to which the statement is made, as a unit or entity in the logic of that system. . . . So soon as there is any system, any organization, with a logic of its own, just so soon must there be some constants, some reference points given, on which to base the logic of the system. Just as the concept "one" in arithmetic is essential to the logical system developed and yet is not one something (e.g., apple or orange, etc.,) so a legal system or any system perhaps must be provided with a basic unit before legal relationships can be devised which will serve the primary purpose of organizing the social facts. The legal person is the unit or entity adopted. For the logic of the system it is just as much a pure "concept" as "one" is in arithmetic. It is just as independent from a human being as one is from an "apple."[8]

Finally, Dewey's famous article on legal personhood nicely articulates the error of the non-Legalist understanding:

> The survey which is undertaken in this paper points to the conclusion that for the purposes of law the conception of "person" is a legal conception; put roughly, "person" signifies what law makes it signify . . . saying that "person" might legally mean whatever the law makes it mean, I am trying to say that "person" might be used simply as a synonym for a right-and-duty-bearing unit. Any such unit would be a person; such a statement

6 Hans Kelsen, *Pure Theory of Law* (1934 German ed, Max Knight tr, University of California Press 1967) 173–174.

7 Jan Klabbers, 'The Concept of Legal Personality' (2005) 11 *Ius Gentium* 35 42.

8 David Derham, 'Theories of Legal Personality' in Leicester Webb (ed), *Legal Personality and Political Pluralism* (Melbourne University Press 1958) 5.

> would be truistic, tautological. Hence it would convey no implications, except that the unit has those rights and duties which the courts find it to have. *What "person" signifies in popular speech, or in psychology, or in philosophy or morals, would be as irrelevant, to employ an exaggerated simile, as it would be to argue that because a wine is called "dry" it has the properties of dry solids.*[9]

Naffine herself summarizes Legalism in a way that contrasts nicely with the Realist understanding we soon turn to.

> In this orthodox and analytical account of law's person, the defining attributes of a being—its capacity to think or feel or its sanctity—have no direct bearing on whether it has legal personality because the legal person is a construct of law. The defining characteristic of law's construct is the formal capacity to bear rights and duties. This does not depend on the supposed essential or even inessential attributes of the being to whom the construct is applied. Rather it depends on, and is formed for, specifically legal purposes. Law's person is not intended to mirror nature, however that is understood. Nor should it be. Instead the legal person is better understood as a piece of pure legal artifice, a clever device of law, law's basic unit of analysis.[10]
>
> Legal persons, thus considered, are legal abstractions. They consist of purely legal norms. Legal persons do not have the innate capacity to reason nor do they possess human souls; they do not have arms and legs or bodies, a sex or a species.[11]

Relatedly, we find Grear advocating use of the expression "legal entity" instead of "legal person," as the former potentially frees us from the non-legal associations and connotations that come to mind when we use the word "person" in legal contexts.

> [T]he term "legal entity" has an advantage over the term "legal person" in so far as it suggests a semantic aperture arresting—as it were—the inevitability of the convergence between legal subjectivity and humanity, thus usefully applying a semantic brake to the anthropomorphism haunting discussions of law's persons. . . . [T]he term "entity" avoids any intrinsic suggestion of subjectivity in the sense of a self-reflective consciousness and/or other incidents taken to constitute the modern notion of

9 John Dewey, 'The Historic Background of Corporate Legal Personality' (1926) 35(6) *Yale Law Journal* 655, 656. My emphasis.
10 Naffine, *Law's Meaning*, 20–21.
11 ibid, 82.

the subject/ person (whether human, animal, intelligent machine) more broadly understood.[12]

Realism Defined

As used by Naffine, Realism (or in some of her writings, "Metaphysical Realism") serves as an umbrella term, capturing different variants. What they have in common, though, is that contrary to the Legalist, person is *not* a mere empty slot, with no connection to non-legal conceptions of persons.

> [The Realist's] conception of the human should be replicated in law's conception of its person, but tend not to defend the need for such a conceptual link. The link is taken as a given: the essential human person should correspond to what is taken to be the essential legal person. Law should mirror life when it makes a legal person. It should try to capture the essence of a personal being when it turns someone into a rights-and-duty bearing entity: a legal person. Law should try to match or capture the nature or quality of life when it personifies.[13]

Importantly, different versions of Realism turn on which aspect of the *non-legal person* one chooses to focus. Thus, in her book *Law's Meaning of Life: Philosophy, Religion, Darwin and the Legal Person*, she distinguishes various accounts of that essential feature, distinguishing those who view the essence of humanity as residing in our ability to reason, versus a religious conception where we are essentially ensouled creatures unique in having immortal souls, versus yet another version according to which we are essentially natural beings who exist in continuum with the rest of the animal kingdom, with the salient quality being our sentience.[14]

> By contrast, the Rationalists, Religionists and Naturalists all believe that the legal person is an expression of some important defining attribute of human nature and therefore it is important to go beyond law to work out what that nature is. They believe that law should find and reflect its subject which exists beyond law (and is variously defined by other disciplines, by philosophy or religion or science). They believe that law finds its subject beyond the legal realm and that law is to be judged by its success at finding and rendering this subject faithfully. The set of beliefs which determines

12 Grear, 'Law's Entities,' 81.

13 ibid, 79.

14 Naffine singles out Finnis as a paradigmatic example of the religious conception, where our rights derive from our nature as beings endowed with a soul and thus as an example of the natural law tradition stemming ultimately from Aquinas. See, for example, John Finnis, *Natural Law and Natural Rights* (Clarendon Press 1980) Chapter 8.

> the extra-legal nature of law's subject will depend on where one goes, in a disciplinary sense, to find out about one's subject.[15]
>
> They seek to divine and give expression to that which makes us truly human—across time and place and culture. They tend also to be essentialists in their thinking about the legal person in that they treat it as a concept with a natural, stable and essential basis (in the human), and which naturally caters for and matches up with this human. They tend not to treat it as an artificial legal concept comprising abstract rights and duties, whose content depends on the variable decisions of law-makers.[16]

Anna Grear refines and develops Naffine's analysis of the various Realist views, elaborating on the different ways in which qualities of persons in the non-legal sense are held to be essential to construing person in the legal sense. *Viz*:

> Naffine, as already noted, also identifies two templates of the "natural person" in law: P2 (P-E2) and P3 (P-E3). Both these constructs presuppose a "natural human being" as the substrate of the "person" in law and rest on operative suppositions concerning "an intrinsic" human nature. P-E2 theories focus most directly, according to Naffine, upon the simple "natural condition of being human" in law. This "human" is legally defined—and rests upon certain biological and metaphysical assumptions concerning the meaning of being "human," or counting as "human." The position presupposes a living human being and a human being becomes a legal being in the P-E2 sense "at birth, which is also legally defined, and stops being a legal person at whole brain death, legally defined." As Naffine points out, in such accounts, legal rights are seen as a mere augmentation of what are taken to be innate moral attributions of "natural" persons.[17]

Grear refines and develops Naffine's analysis of the various Realist views and is keen to focus on a particular version of Realism, one which points to the historical paradigm of personhood, which plays a key role in establishing the historic patterns I alluded to in the Introduction. Here, it is the conception of a person as essentially a rational being, where that is supposed to call to mind the autonomous individual who is fully mentally and legally competent, not reliant on anyone, nor controlled by their emotions. It is this being who is the true possessor of rights. Taken together all of these traits point to the paradigm of a person—the adult male.[18]

15 Naffine, *Law's Meaning*, 22.
16 ibid, 70.
17 Grear, 'Law's Entities,' 12.
18 ibid, 14.

We've now seen contrasting Realist and Legalist understandings of "person." As I quoted Naffine at the start of this chapter, "The largest dispute, the biggest intellectual and moral battle, is between those who say that law does not and should not operate with a natural conception of the person and those who say that it does and should."[19]

With such a long-running debate about the meaning of person already in place, it makes sense to ask how one can advance the discussion, beyond simply repeating what each side holds to be true. Here, I hope to make progress, on behalf of the Legalist, by taking a page out of debates about the rationality of theory change found in the philosophy of science. On one influential view, what makes one scientific theory a better theory than one that came before it is that the later theory is able to not only capture all the data that earlier theory missed but is also, crucially, able to explain how the earlier theory seemed to get so much correct, even though (from the standpoint of the later theory), the earlier theory is literally false. General and special theories of relativity not only explain, for instance, why Newtonian theories of space, time, and gravity *fail* to capture the behavior of bodies at tremendously high speeds or in presence of massive objects; Einstein's theories also explain why Newtonian physics got as much correct as it did, even though that theory is, in point of fact, false.[20]

Likewise, I will try to accommodate as many insights as possible from the Realist side on how the law makes use of "person," but without granting that Realism is correct in its analysis of what "person" means in law. For it would be foolish to deny that the law does often operate with substantive assumptions about the nature of persons in legal reasoning. And there are core or shared assumptions about what that nature is, as the Realist alleges. Yet I think it still possible to accept that there are shared assumptions and commitments, without concluding that those shared assumptions are the ones the Realist focuses on. The shared assumptions, instead, put starkly, are simply that whenever "person" is used in law, we have an instance of something that is the bearer of rights and duties.

So although the following pages will take us far afield strict legal theory, we will end up in a familiar place at the end—the Legalist contention that personhood is open-ended enough to accept a wide range of entities as legal persons, which will include nature, as one would expect. It is the mask, in a sense, that matters, not what lies behind that mask. Along the way, I hope to do more than simply repeat the slogans of Legalism, instead showing that the theory is quite plausible, especially in how it is able to accommodate the undeniable insights of Realism. Again, like a good successor theory in the

19 Naffine, *Law's Meaning*, 20.

20 Here, I draw upon the work of Jay F. Rosenberg on the rationality of scientific theorizing, as found in such papers as his 'Comparing the Incommensurable: Another Look at Convergent Realism' (1988) 54(3) *Philosophical Studies* 163.

philosophy of science, a later theory is able to explain the successes and apparent plausibility of the earlier theories they seek to replace.

Naffine's Objection to Legalism

Naffine herself, though sympathetic to the creative possibilities the Legalist strategy offers, worries that the ways in which the law actually operates shows a fidelity to the Realist position.[21] Thus, according to Naffine, attempts to justify the use of Legalism by appeal to the idea that legal language is, and should be, isolated from the non-legal fails to reckon with the fact that law often does use "person" in ways that *do* involve substantive conceptions of "person," just as the Realist claims.[22] She argues, *inter alia*, that examination of how "person" is used in law reveals that jurists are not free to simply use that term in whatever way they wish. Again, that is, Naffine suggests that although, in theory, "person" may be used in open, unrestricted ways as the Legalist suggests, in point of practice, things are not as simple.

> The legal resistance to the personification of animals strongly suggests that the term *person* is not in fact a slot that fits anyone or anything but rather a slot essentially designed for human beings because they are thought to possess a certain moral status.[23]

Law does often operate with substantive commitments about the nature of persons, often tying the notion of a legal person to a biological or philosophical conception of what human beings are. Despite this acknowledgment, I believe we can still make the case that Legalism better captures the *actual* use of person in the law. I will attempt this through what I will call a *functional-legalist analysis* of the use of *person* in legal language and reasoning. To do so, I will rely on a few methodological constraints, explained later, as well as arguments based on analytic philosophical arguments about the proper way to understand the meaning of terms, what has been dubbed by various authors as "functional role semantics."[24]

21 Naffine, *Law's Meaning*, 367.

22 Particularly in cases involving fetuses and persons in permanent vegetative states, where it is often contested to what extent, if any, those beings have rights, and if they do, what exactly those rights would be. See, for instance, N Naffine, 'Legal Personality and the Natural World: On the Persistence of the Human Measure of Value' (2012) 3 *Journal of Human Rights and the Environment* 68.

23 Naffine, *Law's Meaning*, 356. In subsequent chapters, I will examine arguments that nature has rights because it, too, has intrinsic value and thus a moral status appropriate for rights and personhood.

24 Sometimes referred to as "conceptual role semantics" or "inferential semantics." See Wilfrid Sellars, 'Meaning as Functional Classification (A Perspective on the Relation of Syntax to Semantics)' (1974) 27 *Synthese* 417.

First, though, to be fair, Naffine is surely correct that there are numerous examples of the law relying on substantive conceptions of personhood in a way that gives a nod to Realism. For even if courts indicate that they are not in the business of settling metaphysical or philosophical questions over what is the true essence of persons (or, alternatively, human beings), there are situations where their commitments to particular conceptions of what is essential to persons is on display. How else, we might ask, are we to resolve disputes about the permissibility of abortion, or if life support should be ended for someone in a persistent vegetative state, than by taking a stand on whether the fetus is a person, or whether someone in a permanent coma remains the person they were and thus has the rights they did when they were awake and rational?[25]

After all, if you believe the essence of a person is their rational capacities, you may argue that the comatose patient no longer counts as the same person they were pre-coma. But if you think the essence of personhood is to be an ensouled, living human being, you very well may argue that that very person survives, only now in a coma. If you think simply being a living member of the human species is sufficient for personhood, you might hold that the fetus counts, morally and legally, as person. Thus, apparently contrary to the Legalist, one won't find "person" in legal contexts devoid of substantive, non-legal commitments.

In that sense, Naffine is certainly correct. However, as I argue later, the Legalist need not deny at all that the law often operates with a "thick" conception of person, as opposed to viewing it as an empty slot devoid of any substantive commitments. The key will be to show, as I intend to do, that this is not a problem for the Legalist. That is where I make use of contemporary functionalist accounts in philosophy of language and philosophy of mind.

As a preliminary point, however, it is important to note a way in which Naffine's focus on the substantive commitments that often lay behind use of "person" actually doesn't go far enough in examining how that term is used. Recognizing *how* wide ranging the use of "person" is will actually help the Legalist position. That is to say, if we look at *all* of the ways that "person" is used in law, we find an extremely wide range.[26] Making use not only of

25 We also find extreme cases, such as of the conjoined twins Rosie and Gracie, *Re A (conjoined twins)* [2001] WLR 480, where Rosie had little brain function and seemingly depended on Gracie for survival, where the question before the court was whether to separate them so as allow Gracie to live at the cost of Rosie's death. Re A (Children) (Conjoined Twins: Surgical Separation) [2001] Fam 147 Court of Appeal Ward, Brooke and Robert Walker LJJ <https://www.mentalhealthlaw.co.uk/media/Re_A_(Conjoined_Twins)_(2001)_Fam_147_report.pdf>

26 Naffine is not unaware, of course, of the other uses. Still, if one is going to make a point about legal usage, as she does, then one ought to be explicit about how wide ranging that usage is.

domestic law but also of international law, we find that *all* of the following have been considered as legal persons: individual human beings, families, Indigenous tribes, corporations, partnerships, LLCs, trusts, governments, States, municipalities, universities, associations, cooperatives, religious organizations, international organizations, NGOs, if not more.

Rhetorically, with such a varied list, it is easy to understand the temptation of the Legalist view that *anything* in principle can serve as a person in law, as it seems that nearly anything *has*.[27] Such a wide range of uses lends at least some credence to the Legalist view that anything can in theory be a person in the eyes of the law. Now that is not exactly an airtight argument, as a range of uses doesn't entail that *anything* could be, nor does the fact that some legislature or juristic body has decreed some entity to be a person entail that that entity does, in point of practice, *actually* have a robust set of rights (and/or duties). But it does at least speak to the prima facie openness of the term, suggesting that there is conceptual room for including nature (defined as you like) as also a suitable candidate.

Functionalist Reply to Naffine, Part I

Let us start with a preliminary examination of choices we face at this early stage. Whatever legal powers might be entailed by the granting of personhood to individuals, States, NGOs, LLCs, trusts, etc., and whatever range of rights they might be able to hold in virtue of being deemed *persons*, it is at least the case that the law *does* recognize an extremely wide range of entities as legal persons. Thus, we should ask about the use of "person" and how that extremely broad domain (*extension*) of the term can be a guide to the *intension* of the term.[28]

On one view, "person" can be understood as merely a homonym across these different realms. Corporation, States, NGOs, and human beings are persons, but there is nothing shared in the meaning of the term across those uses. Again, there is no shared meaning across that wide extension, that wide range of things that are deemed legal persons. On this *homonymous* reading, there is nothing more than a shared sound or spelling across the wide range of "person," much as we might say "bank" as a place by the side of a river

27 And that doesn't even include all the emerging cases in which nature itself has been encoded in law as a person, as I discuss later. I leave out mention of those cases so as to avoid begging the question on behalf of RoN advocates.

28 By "extension" I mean the range or set of things that a term applies to. This is in contrast to the "intension" of a term which speaks to the meaning of that term. The *intension* of orange, for instance, would be something like "a colour that is between the colours red and yellow." The extension would be the set of things we call orange—certain fruits, the fur on an orangutan, some people's hair, etc.

has nothing in common with "bank" as an institution where financial transactions occur. Let us call that the *Nominalist* view.[29]

On the other extreme, one might wish to claim that there is a single, univocal meaning of "person" across that wide domain. That when we speak of "person," we are pointing to something shared deeply across "person" *qua* State, *qua* NGO, *qua* corporation, etc.—an essence, if you will. On the former, Nominalist view, there is only the term, the name, that is in common. On the latter, Platonist view, as I will call it, personhood in all cases points to something identical, where the legal-philosophical task involves finding the essence of that shared meaning, that which unites the variety of uses across different realms of law. The Realist view I've described and documented earlier clearly aligns with this Platonist view; both are searching for an essential understanding of "person" to underwrite all uses of that term in law.[30]

I wish to suggest a third way, one that will ultimately be put in service of the Legalist view. It is based on what Aristotle calls *pros hen homonymy,* as used in his *Metaphysics*.

Aristotle's early work, *Categories,* presents what he takes to be an exhaustive list of the kinds of being or entities there are—including substances and properties, relations, places, and number. Given such a list, one may be inclined to ask what all those different kinds of beings or entities have in common, in virtue of which they *are* beings. As a student of Plato, one would expect Aristotle to be searching for the Form of Being, under which all those different kinds of beings can be found but which nevertheless has some unifying essence that connects them all. In more familiar Aristotelian language, this would be to ask after the *genus* that all the different varieties of being fall under—where, say, a color, would be defined by reference to the *genus* of quality and the *differentia* that capture it as a color, as opposed to a shape or sound. And if we look long and hard enough, we will encounter the ultimate genus, being, which connects them all via something, an essence, all these radically different kinds of being have in common.[31]

29 Here, I am deliberately calling to mind the famous Problem of Universals and seeking to show how a debate over the meaning of "person" replicates that philosophical puzzle that started with Plato and continues to this day. For more on the Problem of Universals itself, see Eric M Rubenstein and Mary MacLeod, 'Universals' *The Internet Encyclopedia of Philosophy* <https://iep.utm.edu/universa>

30 Perhaps the most famous (and familiar) of Plato's examples is his investigation of justice in the Republic. What is in common, or shared, for instance, between a just person, a just act, and a just city? What is the essence of justice that these things all have a 'share' of? Plato, *Republic* (CDC Reeve ed, GMA Grube tr, Hackett Publishing Company 1992).

31 Employing the terminology I've used, a Nominalist would say there is nothing shared or in common across that wide range of beings that makes them beings. We simply use that term "being" in multiple unrelated ways. That is, put differently, the homonymous strategy. A Platonist would insist that there is a Form of Being that all the different kinds of beings

But that would be a mistake. Radically simplifying Aristotle's response, he argues that, in fact, there is no overarching genus to be found. All we have are particular kinds of beings—substances, colors, relations, etc. Nevertheless, Aristotle holds that we can understand how all of those things count as beings. Not by finding some single essence they all have in common. Instead, we find his novel solution in what he calls *pros hen homonymy*. For though there is no single meaning across the different kinds of beings, they are also *not* mere homonyms. Instead, all beings are to be understood as beings in virtue of their *relation to a primary type of being*, which for Aristotle is the being of individual substances. What matters is finding that primary being or meaning and seeing how all others are related to that primary example.

The often-cited example of heath is helpful here: various healthy things are called healthy because of the relation to "health," where the meaning of health is the core, and all other things called healthy are related to that core meaning.

"In this case, a regimen, urine, and Socrates are all called 'healthy' not because they stand under some one genus, namely healthy things, *but instead because they all bear some relation to health*."[32]

Let us now apply Aristotle's strategy as a Legalist response to Naffine's worries. On that application, we need not despair simply because "person" is used in such a wide range of contexts, for we can still seek core meaning to "person" and then see how the various uses of "person" in different legal contexts can then be shown to be examples of that *pros hen homonym* phenomenon. Again, that "person" can attach to such wildly different entities entails neither that there is *no* shared meaning, nor does it entail that there is a *single* univocal meaning, as the Realist would have it.

Let me repeat: We are not forced to hold that there is univocal meaning through the variety of uses and contexts where "person" is used, *ala* the Realist/Platonist. Nor need we give up the attempt to find a common meaning as the Nominalist might have us do. Instead, on the Aristotelian understanding I'm suggesting, there could be a core or focal meaning, and the disparate uses of "person" are all linked to that core or focal meaning, without, again, having to hold that the uses are synonymous.

If we can locate such a core or focal meaning, and explain how the different uses all connect, in the variety of ways to that core meaning, we will have found a way to begin to appreciate the power of the Legalist strategy. Namely, that while law considers many different kinds of things to be persons, it is not because there is some singular, deep metaphysical or

partake in, which in turn makes them beings. That would be the Platonist account, applied to *being*, and thus paralleling the Realist strategy for *person*.

32 Paul Studtmann, my emphasis. 'Aristotle's Categories' *Stanford Encyclopedia of Philosophy* <https://plato.stanford.edu/entries/aristotle-categories/>

philosophical conception of what it means to be a person (in an extra-legal sense) that grounds all those uses. Here, we can imagine the Legalist saying that, in fact, it is the assumption that there must be some such philosophical grounding about what it is to be a person that (mistakenly) leads many to think there must be some substantive singular essence or commitment based on moral notions of persons and rights about the nature of beings suited to being legal persons.

On the other hand, the Legalist will insist that their view frees us from that dead-end pursuit, while the alternative—finding a core or focal meaning of "person"—can help us see that while "person" is open ended, it is not a free-for-all in any absurd sense. That core or focal meaning would get us to the heart of what it is to be a legal person. Explaining that core meaning emerges as the central task, though it will not be a surprise that the forthcoming answer will be the familiar one documented earlier—to be the bearer of rights and duties.

Finding that core meaning of "person" that unites the various uses will help remind us of what the point is in talking about *persons* in the first place. I'm suggesting, again, that the *pros hen homonymy* strategy allows us to avoid the pitfalls of Platonism/Realism, without despair that "person" is being used in a problematic free-for-all manner. For there *is* a unity to be found in the uses of "person." It just isn't the unity that the Realist thinks it is.

Functionalist Reply to Naffine, Part II

We return now to the central question: What is the core meaning of "person"? Here, a few methodological constraints should be mentioned. As I see it, an adequate account of this focal meaning should *begin* by being as metaphysically neutral as possible. Given that few jurists see their job as solving philosophical debates over the essence of humanity, and to avoid begging any questions, a theory of "person" in the Aristotelian model I'm suggesting should cover as much of legal usage of "person" as possible, with as few initial theoretical commitments as possible. Relatedly, I will try to refrain, as much as possible, from any *normative* claims about how we should understand "person." In hopes of not begging any questions on behalf of the Legalist who might want to apply "person" to nature, I aim to free my account of as many presuppositions about how the "person" *should* be used as possible.

Using what has been variously described as functional role semantics, conceptual role semantics, or even as befitting perhaps one of the versions from Wittgenstein, meaning as "use," we can account for the meaning of a term by reference to the role it plays in language as a whole. The meaning of terms, more precisely, is determined by the role they play in inferences and reasoning. What role in reasoning then does the term "person" play in the various legal contexts it is used? Making use of subscripts, we can talk about how "person" is used in legal contexts of corporations, or contexts where it is used to speak of States, or for NGOs, etc. We can ask, that is, what is the functional role of "person" in these various cases? And more pointedly, in searching for our core or focal meaning, what is the common role found throughout all of these uses?

"$\text{Person}_{\text{corporation}}$"
"$\text{Person}_{\text{church}}$"
"$\text{Person}_{\text{state}}$"
"$\text{Person}_{\text{individuals}}$"

The straightforward answer, I believe, is that the core meaning in these various uses is that "person" is used to track and distribute rights and duties. The functional role of "person," in other words, is to track and distribute rights and duties—who (or what) has rights and duties, which rights and duties, when those rights are in effect, when they cease to be, etc. The role that "person" plays in law, in other words, is to assign and track at a given time, and over time, which rights and duties the subject in question has. That, I claim, is the core or focal meaning of "person" in law. All uses of person, in the various contexts it is used, are variations on the particular sets of duties and rights that are appropriately assigned at a given time and how they operate across time. The different contexts in which they are used gives us the variety we'd expect when we find "person" used in international law, or domestic law, in corporate law, in criminal law, etc.

When we begin to ask about the particular rights that a given type of entity has, we begin the process of individuating it, and distinguishing it from other kinds of entities, ones that have different sets of rights. The rights that a ship has will differ, of course, from the rights that an infant has, which will differ from the rights that a competent adult has, which will differ from the rights that a corporation has. The point being that in speaking earlier about particular types of persons—indicated by the subscripts—we are looking at different clusters of rights and duties that are appropriate for each kind of entity, each kind of person.

Despite those important differences in how rights and duties are assigned, and in how they operate at a time and over time, there is still something in common among those varied uses of "person." Thus, we need not despair of that term being merely homonymous, nor must we find a single "Platonic" meaning across those uses. "Person" simply functions as a way of assigning rights and duties. And, again, the specific meaning and assignments of rights and duties will vary from context to context. They all have as their core meaning that *generic functional role* of assigning and tracking rights and duties.[33]

33 It would be an interesting project to ask the same question about whether there is a focal meaning for "rights," which after Hohfeld's famous analysis forces us to recognize that "rights" itself has a wide range of meanings, from privileges to liberties to permissions to

All other uses of "person" get their meaning in virtue of their connection with that core meaning of bearing the rights and duties, while the differences among the multitude of "person" simply points to the different sets of rights and duties and how they function differently for each kind of legal person.

The advantage, as I see it, of this strategy is not just that it finds a path between Nominalism and Platonism, as I've used those terms, but it allows us to see that in important ways, Naffine is obviously correct that "person" often is used with substantive philosophical or metaphysical commitments. It would be foolish to deny that. But we can acknowledge that *and* still find a common thread that connects all the varied uses of "person." Crucially, this lets us see that the Legalist is surely also onto something. To put it simply, when we use "person," we are singling out something that is the bearer of rights and duties. That, to use the terminology of earlier, is the functional role of "person." The variety of contexts where "person" is used will be where we find *different* rights and duties assigned. The notion of being a bearer of rights and duties simply indicates the primary, core, or focal meaning of the term, to which all others are related.

Now even if this is true, that doesn't settle the question of whether *nature* is the type of entity that can be the bearer of rights and duties. Addressing that important question will be the task in subsequent chapters. But what we've seen so far is that we can make sense of the varied uses of "person" by showing that they do have something in common, though what they have in common doesn't involve any substantive metaphysical (i.e. Realist) commitments. That, I think, is what the Legalist is after. And I believe the *pros hen homonymy* strategy, along with the function role semantics I'm drawing on, allows us to see that.

Functionalist Reply to Naffine, Part III

Now I wouldn't be surprised if the earlier example fails to convince Naffine or others we've discussed. For they might understandably object that the earlier example looks like a sleight of hand in that it fails to fully grapple with the substantive commitments the law often *does* use when it comes to "person," especially in cases of fetuses, comatose patients, etc. Perhaps even worse, it fails to acknowledge that the way *person* manifests itself in law often reveals a bias towards a particular kind of person, namely, the white male.[34] In order to try and rebut that worry, I bring yet another philosophical

authority. Wesley Newcomb Hohfeld, 'Some Fundamental Legal Conceptions as Applied in Judicial Reasoning' (1913) 23 *Yale Law Journal* 16.

34 See, for instance, Grear on the gendered patterning of "person" in a series of papers and books, including Anna Grear, *Redirecting Human Rights: Facing the Challenge of Corporate Legal Humanity* (Palgrave Macmillan 2010).

consideration to bear on the subject—this time, functionalism as it appears in contemporary analytic philosophy of mind.

One of the most famous papers of 20th-century analytic philosophy is David Lewis's "Mad Pain and Martian Pain," which asks us to imagine various scenarios where we are confronted by unfamiliar beings, call them Martians, who exhibit behavioral symptoms of being in pain.[35] In such a case, though they look to be in pain, were we to examine them, we wouldn't expect their physiology to look anything like ours. They could be in pain even though their biological states are radically different from what ours look like when we are in pain. What then underlies their being in pain if it isn't their shared physiology with us? After all, they might not even have a biology, as perhaps they are mere chemical beings or beings made of substances we don't recognize or understand how they work.

Lewis's answer is that what makes for a Martian being in pain, as opposed to being happy, has to do with how the input of the world's stimuli are mediated by whatever internal state the Martian is in, one that manifests itself (as output) in pain behavior. The pain wouldn't be describable by reference to human anatomy. Instead, suggests Lewis, the Martian is in pain because it is in a particular "functional state," i.e. an internal state of its anatomy that acts as a function, in the fairly literal mathematical meaning, between input and output. Pain, then, could exist in both humans and Martians without any shared physical similarities. The pain would be, in a phrase, a particular functional state, one that differs from other functional states that mediate different inputs and outputs. Hence, the development of *functionalism* in the philosophy of mind.

Put slightly differently, every functional state of mind is instantiated in *some* particular physical medium (as far as we know). But that does not warrant the conclusion that the mental state of pain is identical to only *one particular physical state*, say, the physical pain state that humans have. Instead, pain is a functional state that can be instantiated in variously diverse physical states, such as in humans, in animals, in Martians, etc.

The connection with Naffine and others that I'd want to draw is that in a similar way to how we understand pain as a functional state, we can now understand the core or focal meaning of person as itself a kind of functional state. The core meaning behind different uses of "person" we've uncovered would be what different particular manifestations of "person" have in law: how "person" is *instantiated in different legal contexts*, with different clusters of rights and duties as we move from one context to another. Just as pain would be a functional state with different particular manifestations or instantiations in different physical states, *person* indicates a cluster of rights and duties—that is the function of *person,* while different persons will be

35 David Lewis, 'Mad Pain and Martian Pain' in Ned Joel Block (ed), *Readings in the Philosophy of Psychology* (Cambridge 1978) 216.

indicated or individuate, or shall we say, manifested differently as we move from one legal context to another.

So to reply to Naffine's worries about those patterns of rights and duties that have typically had the adult male as their paradigm, we can say the following. Historically, yes, "person" has often functioned in a way that gives rights and duties first and foremost to property owning white men, and the assignment of rights to other beings depends on the extent to which they are like those white men. But for "person" itself, the function would be to simply track and distribute rights and duties. In corporate law, in contrast, "person" would be instantiated to have different rights and duties. For instance, corporations have the right to own property and, in some sense, also to be owned, whereas as "persons" in the noncorporate realm are typically beings that are not able to be owned. That there would be different rights and duties across the multitude of legal contexts should not be surprising. But like the semantic account given earlier about the core meaning of "person," we need not be concerned that person is instantiated in different ways—that is, with different rights and duties—across different realms. That is as it should be. Just the same, what they all have in common is that same function—of mediating and distributing rights and duties in those particular contexts.

The core idea, again, is that when speaking of "person" in the legal sense, that usage is perfectly compatible with a range of particular instantiations in different legal contexts. What matters, to refer back to the terminology of Realism and Legalism, is that that legal person is a subject of law, one with rights and duties. The personhood may be tied to a corporation in one context, to a university in another, to a physical human being. But "person" in the legal sense is flexible enough to accommodate all of those particular instantiations.

This, I take it, has been a very long-winded way of articulating what is meant by that passage from Naffine herself, quoted earlier in this chapter on behalf of the Legalist:

> In this orthodox and analytical account of law's person, the defining attributes of a being—its capacity to think or feel or its sanctity—have no direct bearing on whether it has legal personality because the legal person is a construct of law. The defining characteristic of law's construct is the formal capacity to bear rights and duties. This does not depend on the supposed essential or even inessential attributes of the being to whom the construct is applied. Rather it depends on, and is formed for, specifically legal purposes. Law's person is not intended to mirror nature, however that is understood. Nor should it be. Instead the legal person is better understood as a piece of pure legal artifice, a clever device of law, law's basic unit of analysis.[36]

36 Naffine, *Law's Meaning*, 20–21.

What I hope to have accomplished in the earlier forays into the history of philosophy is to find a way to advance the Legalist argument that "person" need not be constrained by any particular philosophical (or otherwise) conception of person outside of the legal context. I've tried to show this by essentially granting all of the data that the Realist has pointed to regarding how the law uses "person," while at the same time showing that the various instantiations of "person," even with their often substantive commitments, do not entail that all uses of "person" have those commitments. What is crucially common across all the uses of "person" in law is simply that there be a subject that has rights and duties. That is the function of legal personhood, and as long as that function is played, keeping track of rights and duties, at a time and over time, we have what it takes for something to be a legal person.

And though it may be overly optimistic, focusing on what the concept of "person" actually implies, to simply be a bearer of rights and duties in the Legalist sense, may serve as a pointed reminder to jurists that personhood is *not and should not* be tied to particular conceptions of human beings. To confront the problem of patterning, as far as legal theory goes, as opposed to social-political theory, perhaps progress can be made by focusing again on the core function of making something or someone a person in the first place. To elevate them such that they become legally visible by having rights and duties. The problem of entrenched patterns that take as their paradigm for persons as white males is not easily solved, of course. I've tried to suggest that there is an understanding of personhood that can be liberating and which offers a way out of that entrenched thinking. For it contains the seeds of possibility—of expanding our conceptions of who the law is for and the possibility of giving personhood to previously considered non-persons. Even, perhaps, to portions of nature beyond human beings.

A Problem for an "Orthodox View"?

Visa Kurki has offered an extended critique of what he calls the "Orthodox View" in legal theory, a critique that the position I've advanced seems vulnerable to.

> According to the Orthodox View, one's being a legal person is the same thing as one's holding of legal rights and/ or bearing of legal duties, or one's capacity to hold such rights and/ or bear such duties. This view is almost always propounded or assumed when legal personhood is discussed. . . . If the Orthodox View's equation of legal personhood and right-holding is assessed in light of Hohfeld's theory, many of the widely held extensional beliefs come into conflict with the Orthodox View: modern theories of rights, which are based on the Hohfeldian analysis, either ascribe rights to entities that are not usually classified as legal persons, such as fetuses and nonhuman animals, or deny rights to entities that are ordinarily classified

as legal persons, such as human children. . . . Slavery is an obvious example. Slaves are very commonly taken to have been legal nonpersons. This would imply according to the Orthodox View that slaves held no rights and bore no duties, and some scholars, such as Hans Kelsen, assert this explicitly. At least as far as duties are concerned, the claim is simply false. The fact that slaveholding societies commonly held slaves legally responsible for their delicts is for slavery scholar Orlando Patterson reason enough to conclude that the "common definition of a slave as someone without legal personhood" is a "fallacy that we can quickly dispose of."[37]

Rather than being a black- and-white affair, legal personhood comes in shades of grey—it is a cluster property. A legal person is not simply a right-holder or a duty-bearer; rather, legal personhood consists of divisible but interconnected incidents of legal personhood.[38]

On the Orthodox View, says Kurki, to be a legal person just is to be the bearer of rights and duties. That is the view I've articulated, and so it seems I must respond to objections from Kurki. In short, Kurki is suggesting that being a legal person is neither necessary nor sufficient for rights and duties. Or at least, in his view, our beliefs about who or what is a person don't consistently align with judgments about who or what has rights and duties. Here, personhood seems not to be logically *necessary* for rights and duties, as we can find historical examples of beings having rights and/or duties despite consistently being denied legal personhood. He gives the example of slaves.

From the other direction, being a legal person doesn't seem *sufficient* for having rights and duties, for though children are legal persons, they are denied a host of rights and duties that fully grown competent adults are given. Again, there is a clash between our judgments about who is a legal person and who has rights and duties.

Kurki concludes the problem with the Orthodox View is its insistence in treating personhood as an all-or-nothing affair: One either is a person with rights and duties or one is not a person and does not have rights or duties. Instead, says Kurki, personhood should be viewed as a cluster concept, comprised of different incidents of personhood, which allows us to distinguish different kinds of personhood, depending on whether at least some of the unspecified number of incidents are present, giving us Passive Incidents of Legal Personhood (such as the right against bodily harm) and Active Incidents of Legal Personhood (such as having positive duties something must fulfill).[39]

So for Kurki, it seems we must distinguish two distinct kinds of personhood—Active and Passive, each with their distinct respective incidents. I have argued

37 V Kurki, *Theory of Legal Personhood* (Oxford University Press 2019) 15.
38 ibid, vii–viii.
39 ibid, 95.

that being a legal personhood *just is* being the bearer of rights/duties, and so one may wonder whether my view is vulnerable to Kurki's critique. I believe it is not.

Holding, as I do, that personhood is simply to be the bearer of rights/duties is consistent with leaving open exactly which rights/duties are appropriate. That is, we can accept with Kurki that being a legal person is shorthand for either Passive or Active personhood, with their respective sets of rights/duties; and whichever type of personhood we are dealing with, we still are talking about something as the bearer of particular rights/duties. Which kind of personhood will depend on what particular rights or duties something has. That seems like an important insight—that the question isn't simply who or what is a person but what rights and duties they have. Yet I believe that is exactly what I've maintained earlier. Different types of persons can be seen as individuated by what their particular "bundle" of rights/duties includes. An infant's bundle will look different than a corporation. If we want to follow Kurki and specify that the infants' bundle makes them merely a passive legal person, so be it; it does indeed help to be clear on how the bundles of rights and duties differ across types of persons. As I see it, that is exactly what I've been trying to articulate.

As far as RoN is concerned, which will be the focus in subsequent chapters, it is clear that the focus will be on the *passive* incidents Kurki discusses—the potential right that nature has to something like bodily integrity, which could ground a claim that in some circumstances, it has suffered a legal harm requiring redress. This is the kind of right, and attendant harm, that is often highlighted by RoN, along with others, such as the right to exist, the right to flourish, the right to regenerate, and the right to evolve—the kinds of rights against harm and destruction.

Thus, I am fine to grant Kurki that we should avoid simply assuming that there is only one form personhood can take. But having specified, using Kurki's analysis, which kinds of rights are in question, we can still stick with the convenient shorthand I've outlined and defended earlier—that to be a person is to have various rights and duties—so long as we are clear, at least implicitly and, better yet, explicitly, about which rights/duties are in play.

Recall the bigger picture at issue here as well. Namely, we are searching for a conception of personhood best suited to RoN. I have suggested, and will continue to develop and defend, a conception of personhood that strips it of biological, metaphysical, and even religious conceptions of humanity. That is in service of defending what I've called a Legalist, as opposed to a Realist, conception of legal personhood. Thinking of persons in the, as it were, stripped-down way I've been exploring serves that goal, offering a conception of personhood that conceives of it as simply a basic notion that allows us to assign and track rights and duties.

Conclusion

Asking what kinds of beings should be considered legal persons is, of course, a vexed question. In coming chapters, I will argue that nature does have what it takes to qualify. But the normative question of whether nature ought to be a legal person, and what rights and duties, if any, it should have, presumes we are clear about the descriptive question of what it is to be a legal person in the first place. To answer that question, I've suggested we look at the wide range of ways in which "person" appears in law. That extremely varied usage does lend at least some credence to the view we've seen in such authors as Kelsen, Derham, and Dewey, to name a few, which suggest that "person" is open-ended enough to serve the needs of law, whatever they may be. Because "person" is nothing more than being the bearer of rights and duties, it is an extremely flexible notion. That is the advantage of Legalism, as I see it. Unlike Realism, which carries heavy metaphysical (or biological or religious) baggage on what could be a person, there is something liberating about having such a pared-down understanding of legal persons.

What I've concluded from that cursory survey of its various uses is that "person" does no more nor less than to simply indicate that we are speaking of a being that has rights and duties, where law is free to stipulate which rights and or duties the person in question has. What has been rejected, on the other hand, is a view whereby law has its hands tied in the use of "person," requiring legal persons to look like or at least resemble in some important manner the nature of human beings. Crucially, as I see it, this understanding implies a flexibility that can be called on when law is wanting to draw on its creative capacities.

Recall the quotations from Blackmun and Douglas we began with. They spoke of a need for law to be able to address developing problems, to be a flexible enough instrument that it can be called on as societies face new dilemmas and new obstacles. They gave voice to a fear that we might be stuck with concepts, rules, and laws ill-suited to developing problems. My hope is that the preceding pages have begun to set the stage for a conception of persons that can underwrite the rights that are needed to address the fundamental threats to nature. In short, that a Legalist view of persons just might be a tool flexible enough to let law develop and to be open to new and creative solutions that are needed now, in a way they perhaps were not needed a hundred years ago.

A quasi-historical point may offer further inspiration for the Legalist strategy of liberating the concept of legal persons from conceptions of the human being. I began this chapter by recalling arguments suggesting the notion of "persona" initially began as meaning simply a theatrical mask presented to an audience to represent a character, where who or what lies under that mask was immaterial. But that conception changed and began to take on significance as a human-centered being, the liberal individual, who comprises the fundamental elements of society, individuals in need of protection from others and protection perhaps even more so from increasingly powerful States.

The important point here is that concepts change their meanings over time. Thus, while "person" may have been the term used through centuries of law, its meaning can and did shift. With the expanding promise and potential of liberal democracies worldwide, it is understandable that humans would become the center of law, with law having at least one of its priorities to protect the rights (variously understood and defined and grounded) of humans. The Universal Declaration of Human Rights in 1948 signaled a high-water mark, full of promise for the protection of humans from various harms and indignities.[40] But like time, the meaning of words marches on. So while it may have been correct to equate "person" and "human being" at one point, there is no reason to think the two are forever linked as equivalent. And I've tried to show how they can come apart. And if they do, that opens the door for "person" to continue to take on new meanings and to be used in new and novel ways, ones suited to the tasks law and society sets for themselves.

Consider the term "gravity," which, for Newton, indicated an ill-defined mysterious force that bound objects large and small together. In the hands of Einstein and the general theory of relativity, "gravity" came to mean something radically different, namely, a literal warping of space-time. It wasn't simply that our beliefs about gravity changed but that the term itself came to mean something different. So, too, for "atom," which, while literally meaning something indivisible, now designates a host of particles bound together by various forces. So, too, for the linkage between geometry and what we later came to realize was traditional build around one form of geometry—Euclidean, while there are other kinds as well. That discovery required us to sever the notion of "geometry" as necessarily implying a Euclidean construal.

Likewise, "person" in the legal sense may have long served as interchangeable with humanity, especially given the concerns and priorities of law. But as law has developed and become more complex, that connection is no longer essential, as we've seen by simply considering the wide range of entities that are now fully fledged legal persons, despite having no connection with humans or any of the traits deemed essential to the latter. Freeing "person" from its early and widespread usage, seeing it has having changed its meaning is to open the possibility that it can change again. What remains to be argued, as I will attempt in subsequent chapters, is that the concept of person is such that it can be employed on behalf of nature and is thus a suitable entity worthy of the elevation that comes with that lofty term, with the possibility of deriving rights for nature—rights that call on us to implement changes in our thinking and behavior.

References

Derham D, 'Theories of Legal Personality' in L Webb (ed), *Legal Personality and Political Pluralism* (Melbourne University Press 1958) 5.

Dewey J, 'The Historic Background of Corporate Legal Personality' (1926) 35(6) *Yale Law Journal* 655.

40 Universal Declaration of Human Rights, *United Nations General Assembly in Paris on 10 December 1948* (General Assembly Resolution 217 A) <https://docs.un.org/en/A/RES/217(III)>

Finnis J, *Natural Law and Natural Rights* (Clarendon Press 1980).

Finnis J, 'Natural Law Theories' in *Stanford Encyclopedia of Philosophy* <https://plato.stanford.edu/entries/natural-law-theories/>

Grear A, 'Human Rights—Human Bodies? Some Reflections on Corporate Human Rights Distortion, the Legal Subject, Embodiment and Human Rights Theory' (2006) 17 *Law Critique* 171 <https://papers.ssrn.com/sol3/papers.cfm?abstract_id=1825128>

Grear A, *Redirecting Human Rights: Facing the Challenge of Corporate Legal Humanity* (Palgrave Macmillan 2010).

Grear A, 'Law's Entities: Complexity, Plasticity and Justice' (2013) 4(1) *Jurisprudence* 76 <https://www.tandfonline.com/doi/abs/10.5235/20403313.4.1.76>

Hohfeld WN, 'Some Fundamental Legal Conceptions as Applied in Judicial Reasoning' (1913) 23 *Yale Law Journal* 16.

IN RE A (Children) (Conjoined Twins: Surgical Separation) (2001) Fam 147 Court of Appeal Ward, Brooke and Robert Walker LJJ <https://www.mentalhealthlaw.co.uk/media/Re_A_(Conjoined_Twins)_(2001)_Fam_147_report.pdf>

Kelsen H, *General Theory of Law and State* (Harvard University Press 1945).

Kelsen H, *Pure Theory of Law* (1934 German ed, M Knight tr, University of California Press 1967).

Klabbers J, 'The Concept of Legal Personality' (2005) 11 *Ius Gentium* 35.

Kurki V, *Theory of Legal Personhood* (Oxford University Press 2019).

Lawson FH, 'The Creative Use of Legal Concepts' (1957) 32 *NYU Law Review* 909.

Lewis D, 'Mad Pain and Martian Pain' in N Block (ed), *Readings in the Philosophy of Psychology* (Cambridge 1978) 216.

Naffine N, *Law's Meaning of Life: Philosophy, Religion, Darwin and the Legal Person* (Hart Publishing 2009).

Naffine N, 'Legal Personality and the Natural World: On the Persistence of the Human Measure of Value' (2012) 3 *Journal of Human Rights and the Environment* 3.

Plato, *Republic* (CDC Reeve ed, GMA Grube tr, Hackett Publishing Company 1992).

Rosenberg JF, 'Comparing the Incommensurable: Another Look at Convergent Realism' (1988) 54(3) *Philosophical Studies* 163.

Rubenstein E and MacLeod M, 'Universals' *The Internet Encyclopedia of Philosophy*, <https://iep.utm.edu/universa/>

Sellars W, 'Meaning as Functional Classification (A Perspective on the Relation of Syntax to Semantics)' (1974) 27 *Synthese* 417.

Stone C, 'Should Trees Have Standing?—Towards Legal Rights for Natural Objects' (1972) 45 *Southern California Law Review* 488.

Stone C, *Should Trees Have Standing?: Law, Morality, and the Environment* (Oxford University Press 2010).

Studtmann P, 'Aristotle's Categories' *Stanford Encyclopedia of Philosophy* <https://plato.stanford.edu/entries/aristotle-categories/>

Universal Declaration of Human Rights, *United Nations General Assembly in Paris on 10 December 1948* (General Assembly Resolution 217 A) <https://docs.un.org/en/A/RES/217(III)>

Wesley Newcomb Hohfeld, 'Some Fundamental Legal Conceptions as Applied in Judicial Reasoning' (1913) 23 *Yale Law Journal* 16.

Wise S, 'Legal Personhood and the Nonhuman Rights Project' (2010) 17 *Animal Law Review* 1.

Chapter 2

Personhood Without Anthropomorphism

The previous chapter argued that the use of "person" across various legal contexts is best understood as simply indicating something that is the bearer of rights and duties. In that way, I have sought to introduce and defend a Legalist understanding, arguing that the Realist is incorrect in claiming that the proper use of "person" is tied in any deep way to conceptions of persons in non-legal contexts. The concept of the legal person, in other words, is not necessarily reflective of conceptions of persons given by philosophy, biology, or religion.

In separating the legal concept of personhood from the non-legal, I have sought to remove, as it were, any hint of anthropomorphism in the concept of a legal person. Though this is perhaps not how Legalism has historically been presented, arguing that the law's use of "person" has no deep connection with personhood considered in other contexts is, in essence, arguing that the legal person need not have any resemblance to human beings. That corporations, States, LLCs, etc., as explored previously, are just as much legal persons as individual humans clearly entails that the legal person need not be rational, nor sentient, nor ensouled. All that is involved in speaking of a legal person is that it be the bearer of rights and duties. It bears remembering Derham, quoted in the previous chapter, on how *both* the so-called "juristic person" and the "natural person" are creatures of the law:

> In both cases the use of the term "legal person" is as necessary a part of a statement which if true or possibly true asserts, claims, decides, or assumes that a designated entity (whether a man, a group of men, a fund, an idol, or a purpose) is treated . . . as a unit or entity in the logic of that system.[1]

If the earlier analysis is correct, we have in place a preliminary understanding of how "person" is used in the law. The functionalist account I developed,

1 D Derham, 'Theories of Legal Personality' in L Webb (ed), *Legal Personality and Political Pluralism* (Melbourne University Press 1958) 5.

DOI: 10.4324/9781003740186-3

focusing on the core meaning of "person" across different legal contexts, provides an explication. Crucially, though, it is a separate project to ask how "person" *should* be used and which entities are appropriate for that highest level of legal visibility. Ultimately, this aims at the question of whether nature is suitable for personhood. Should nature be given the cluster of rights (and/or duties) befitting something that has the status of a legal person?

Even if the earlier analysis is correct that "person" simply denotes the bearer of rights and duties, as the Legalist maintains, that is *not* tantamount to saying that nature should be such a bearer. For at this point, RoN is struggling against the tide, we might say, for even considering the ever-increasing number of domestic legal contexts where RoN has gotten a foothold, it remains the case that international law does not consider nature a legal subject. Nor is it yet clear, even in those domestic contexts, what legal effect there has been from nature's declaration as legal person.[2] Suggesting that nature *should* be considered a legal person requires further argument.

In this chapter, I will turn a critical eye towards various strategies that have been used to suggest that nature should be viewed as a legal person. As I see it, the central issue is *how* those arguments on behalf of nature have been constructed. Put simply, I will suggest that attempts to establish the legal personhood of nature on Realist foundations or premises are misguided. *If* nature is to be given legal personhood, it should not be based on Realist conceptions and arguments. Though Realist and Legalist advocates of RoN may have the same goals, the *theoretical viability* of their different foundations is of great importance.[3]

In particular, in this chapter, I aim to undermine the first of two broad Realist strategies that have been used to ground the claim that nature should be afforded rights and legal personhood. One strategy can be thought of as seeking to assimilate nature to recognizably human traits. This is the claim that the non-human parts of nature are sufficiently like humans as to warrant the former being given its own rights and personhood. Here, I will suggest this strategy involves an anthropomorphism that simply cannot withstand critical scrutiny. The second strategy, which I examine in the next chapter, critically assesses attempts to ground nature's personhood and rights on moral considerations, particularly the idea that nature has intrinsic value or worth, which in turn provides a suitable foundation for *legal* rights. To be clear, however, the two problems I will point out—problematic anthropomorphism and unconvincing ethical arguments—will be largely schematic. Though it is not

2 Later chapters explore in detail the progress in various domestic cases.

3 Chapter 3 argues for the antecedent. Namely because nature has *interests*, it should be considered a legal person. The present chapter aims to show that, were we to give nature legal personhood, it is best done by drawing on Legalist, not Realist, assumptions about the nature of personhood.

possible to cover all variations of these arguments, should we find reasons to be skeptical of both argument strategies and *also* find a less problematic route via Legalism (as I will provide in Chapter 4), I will consider that an important conclusion to have reached.

Descriptive and Normative Conceptions of Personhood

If Realism sees personhood as intimately connected with human nature (conceived in various ways), it would be no surprise that attempts to make the rest of nature a legal person will do so by making nature seem sufficiently human-like, whether that assimilation is explicitly intended or not. With enough human-like qualities, nature can be said to resemble the paradigmatic legal person, which, again, the Realist conceives as being essentially human. That is the connection between Realism and anthropomorphism I mentioned in passing at the start of this chapter. And once we are aware of the ease of characterizing non-human nature in human terms, we can spot attempts to piggyback nature's having rights onto this assimilation.

In what follows, I will argue that in the quest for making nature into a legal person, nature is often inappropriately and misleadingly made in the image of humans. That is the anthropomorphism (and anthropocentrism) I wish to reject, and it is important to see both how widespread and easy it is to make that (mistaken) connection. In contrast, I recommend rejecting Realist arguments whenever they invoke anthropomorphic considerations. To put it in a more pithy way, we might note the irony that, while much of RoN is motivated by a desire to move past *anthropocentrism* that has been central to environmental law and policy, it fares no better if it does so while embracing *anthropomorphism*. Insisting on the similarities between the non-human and human parts of nature puts us in danger of relying on the same problematic thought that motivated the ancient Greek philosopher Protagoras, namely that "man is the measure of all things."[4] The problem, to be clear, isn't with the idea of nature having rights or being a legal person *per se*. Rather, it is the Realist route to the goal which is problematic. So, too, when it relies on such unwieldy notion as *intrinsic value,* which will be our concern in the next chapter.

Ease of Anthropomorphism

David Hume's *Dialogues on Natural Religion* systematically explores various arguments for God's existence, focusing in particular on the so-called design argument. That argument contends that observation of the natural world's complexity and machine-like features makes it reasonable to infer that the

4 "Man is the measure of all things: of the things that are, that they are, of the things that are not, that they are not." D Bostock, *Plato's Theaetetus* (Oxford University Press 1988) 152a.

world has been created by an intelligent designer. Among the many observations and critiques Hume makes about this argument is one that speaks to the ease and naturalness of seeing the world through the lens of what is familiar to us. It is no surprise, given the high-regard we have for our own powers of creation and design, i.e. our rationality and intelligence, that we are prone to see features of the world as created by something with the vary qualities we value—intelligence.

To help make this point, Hume presents a thought experiment of sorts: a world occupied by (thinking) spiders would understandably spin (pun!) a cosmology and cosmogony that reflected what they both knew first and foremost and what they valued—an infinite spider with infinite web-spinning powers.

> The Brahmins assert, that the world arose from an infinite spider, who spun this whole complicated mass from his bowels, and annihilates afterwards the whole or any part of it, by absorbing it again, and resolving it into his own essence. Here is a species of cosmogony, which appears to us ridiculous; because a spider is a little contemptible animal, whose operations we are never likely to take for a model of the whole universe. But still here is a new species of analogy, even in our globe. And were there a planet wholly inhabited by spiders, (which is very possible,) this inference would there appear as natural and irrefragable as that which in our planet ascribes the origin of all things to design and intelligence.[5]

A world of spiders would see their world as the product of an elaborate spinning being; their world an infinitely complex spiderweb. That cosmology would be just as natural for them as it for us to see our world as machine-like and the creation of an intelligent designer.

Now Hume's point isn't necessarily to criticize this anthropomorphism (*arachnomorphism*?) but to point out how natural it is for us to view the world that way. Thus, with ease, we say things like:

My cat *wants* to go outside.
My dog is *happy* to see me.
That crow *hates* you.
Elephants never *forget*.
The river *wants* to run its own course.
The lawn *wants* to be watered.
The sky *threatens* rain.
The sea is *angry* today.
Nature is *cruel*.

5 David Hume, *Dialogues Concerning Natural Religion* (Clarendon Press 1935), Part VII.

> The dishwasher *hates* it when I forget to put new cleaning pods in.
> The garage door *wants* to avoid crushing object in its path.
> The self-driving car *knows* that there are pedestrians on the sidewalk.
> My car *wants* to be warmed up before I drive it.

My point at this stage of the argument is not a critical one about whether it is appropriate to attribute mental states to things other than humans. In various earlier examples, many will be inclined to accept some attributions of mental states, say to their pets or other animals. Rather, it is simply to point out the ease and frequency with which we do it.[6] In later sections, however, I will engage in critical evaluation. For it will be important to assess the plausibility of such anthropomorphism when we find it employed on behalf of arguments that nature is deserving of rights and personhood because of its human-like qualities, as in this passage from Christopher Stone that I return to later.

> As for the first objection, natural objects *can* communicate their wants (needs) to us, and in ways that are not terribly ambiguous. I am sure I can judge with more certainty and meaningfulness whether and when my lawn wants (needs) water, than the Attorney General can judge whether and when the United States wants (needs) to take an appeal from an adverse judgment by a lower court. The lawn tells me that it wants water by a certain dryness of the blades and soil-immediately obvious to the touch-the appearance of bald spots, yellowing, and a lack of springiness after being walked on.[7]

Layers of Anthropomorphism in Realist Arguments

What I wish to do in this section is show how easy and tempting it can be to argue for nature's rights on Realist assumptions that involve anthropomorphism. As I've said, if your goal as a RoN advocate is to make the case that nature should be considered a legal person, one route to that conclusion would be with a Realist understanding of personhood we've been exploring. Because personhood on the Realist reading is tied essentially to the nature of human beings, one can be expected to demonstrate the similarity or

6 Outside of contexts where rights are at issue is the ongoing reading of nature as having human-like traits, perfectly exhibited in recent scientific literature on *plant intelligence*. For instance, Jennifer Khattar, Paco Calvo, Ina Vandebroek, Camilla Pandolfi, and Farid Dahdouh-Guebas, 'Understanding Interdisciplinary Perspectives of Plant Intelligence: Is it a Matter of Science, Language, or Subjectivity?' (2002) 18 *Journal of Ethnobiology and Ethnomedicine* 41 <https://doi.org/10.1186/s13002-022-00539-3>

7 CD Stone, *Should Trees Have Standing? Law, Morality, and the Environment* (3rd edn, Oxford University Press 2010) 11. Quoted with permission from Oxford Publishing Ltd.

resemblance of nature with one of those essential human features, features that are deemed essential to personhood.[8]

As a rhetorical device, there seems to be empirical evidence that employing such anthropomorphism can actually help foster pro-environmental attitudes.

> Researchers have found anthropomorphism of nature to be positively correlated with pro-environmental attitudes and there is some experimental evidence that manipulating anthropomorphism leads to increases in pro-environmental attitudes.[9]
>
> This reasoning points to the possibility that when an entity is anthropomorphized, it becomes a source of social connection. . . . Because anthropomorphism of nature highlights the similarity between nature and humans, this hypothesis is also consistent with the notion that similarity breeds affiliation.[10]

That does make anthropomorphic considerations plausible as a strategy to increase sympathy with various environmental goals and policies. But sympathy is different from providing solid reasoning. It is the latter I'm concerned with.

As an example of anthropomorphic-based arguments, consider one put forth by Tom Regan on animal rights. To be clear, his arguments are about *moral* rights and *moral* conceptions of persons, not legal. But the form of his reasoning parallels what we may find in arguments for *legal* rights.

Regan's view, in short, is that (at least some) animals are sufficiently like us in that they are experiencing "subjects of a life" and are, therefore, deserving of (moral) rights.

> Animals, it is true, lack many of the abilities humans possess. The can't read, do higher mathematics, build a bookcase, or make *baba ghanoush*. Neither can many human beings, however, and yet we don't (and shouldn't) say that they (these humans) therefore have less inherent value, less of a right to be treated with respect, than do others. It is the *similarities* between those human beings who most clearly, most non-controversially

8 Importantly, as we'll see later, *critics* of extending personhood to nature may themselves still accept that the ground on which one should (or shouldn't) grant personhood to nature turns on how close the resemblance is between nature and human beings.

9 Marc O Williams, Lorraine Whitmarsh, Diarmait Mac and Giolla Chríost, 'The Association Between Anthropomorphism of Nature and Pro-Environmental Variables: A Systematic Review' (2021) 255 *Biological Conservation* 2 <https://doi.org/10.1016/j.biocon.2021.109022>

10 Kim-Pong Tam, Sau-Lai Lee and Melody Manchi Chao, 'Saving Mr. Nature: Anthropomorphism Enhances Connectedness to and Protectiveness Toward Nature' (2013) 49(3) *Journal of Experimental Social Psychology* 514, 515 <https://doi.org/10.1016/j.jesp.2013.02.001>

> have such value (the people reading this, for example), not our differences, that matter most. And the really crucial, the basic similarity it simply this: we are each of us the experiencing subject of a life, a conscious creature having an individual welfare that has importance to us whatever our usefulness to others.[11]

Having such rights on Regan's view means there are limits to what can be done with animals, even if such treatment would maximize overall happiness. Rejecting utilitarian views, that is, about why animals should be treated differently than they presently are, Regan grounds his view in something like a Kantian view of persons, according to which the salient feature deserving of rights is autonomy.[12] Since full autonomy might be limited to only human animals, Regan can be seen as wanting to keep the spirit of a Kantian anti-utilitarian conception of how others should be treated but weakening it sufficiently to allow many non-human animals to be considered worthy of having rights. Crucially, again, it is the similarity with us that matters.

A similar strategy of looking for similarity is found in Peter Singer, though for Singer, the salient feature we share with animals is our *sentience*. That is the relevant feature determining whether or not something is deserving of special treatment.[13] So while Singer disagrees with Regan on which feature of humans is important to find replicated in the animal world, both rely on a similar type of argument, one that Beaudry has usefully dubbed "assimilationist."[14]

These arguments have the following form, where one is attempting to argue that some class of beings should have rights that presently they do not. We begin by pointing to a class of beings that *do* have such rights, along with certain properties. We then reason that because the beings in question are sufficiently similar to that privileged class, we should grant rights to the contested class. We might note, say, that white human males are rational and are deserving of rights. Then, because non-white women also are rational, they, too, are deserving of rights. It is the shared property, or at least the similarity between the two groups, that warrants the extension of rights from the first class to the second.

11 Tom Regan, 'The Case for Animal Rights' (1985) 8(4) *Human Studies* 389, 394.

12 See Thomas E Hill, 'The Kantian Conception of Autonomy' in John Philip Christman (ed), *The Inner Citadel: Essays on Individual Autonomy* (Oxford University Press 1989) 91.

13 I say "special treatment" instead of "rights," as Singer's utilitarianism in theory seems to allow anything to be done to even sentient beings if it would maximize happiness. That can be seen as antithetical to a conception of rights, which put limits on what can be done even if it maximizes happiness.

14 Jonas-Sébastien Beaudry, 'From Autonomy to Habeas Corpus: Animal Rights Activists Take the Parameters of Legal Personhood to Court' (2016) 1 *Global Journal of Animal Law* 3 <https://journal.fi/gjal/article/download/148714/94849/358299>

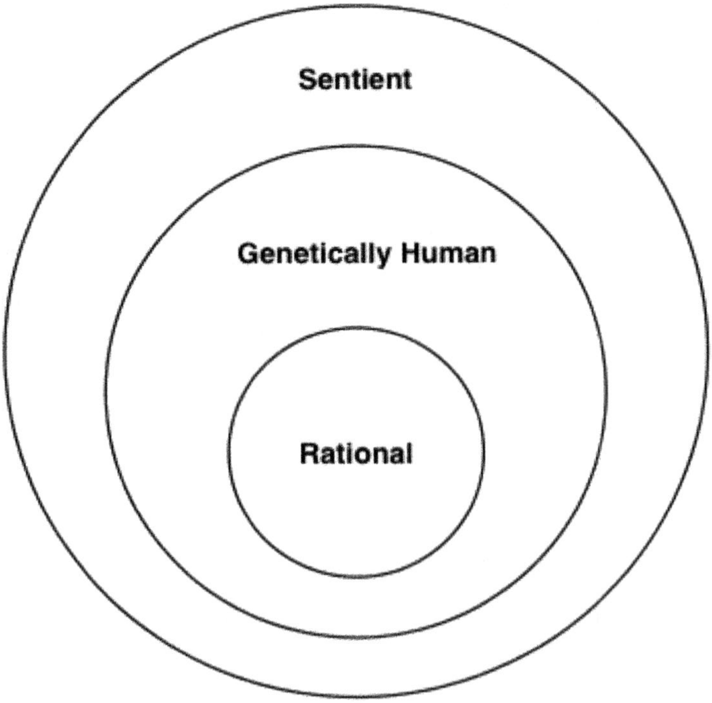

Figure 2.1 "Realism"

What I'm particularly interested is the use of such assimilationist arguments by Realist RoN advocates for nature having legal personhood and legal rights. On this logic, the key move involves finding relevant similarities between nature and human beings, where for the Realist, the feature you select will be what you see as essential to human nature. As for which traits are significant, we can imagine an expanding series of concentric circles, with each circle encompassing a large domain of beings that have the quality in question, getting more and more general and inclusive as we expand the circles.

On this way of viewing Realism, the most restrictive feature on which personhood depends would be being a rational human being.[15] The class of

15 Providing a definition of "rational" is notoriously difficult. An important primer on the challenge and how to meet it can be found in Daniel Dennett, 'Jonathan Bennett's *Rationality*' in Eric Schliesser (ed), *Ten Neglected Classics of Philosophy* (Oxford Academic Press 2016) <https://doi.org/10.1093/acprof:oso/9780199928903.003.0010>

genetically human beings, on the other hand, would be slightly more inclusive, encompassing both rational humans as well as human infants, humans in permanent vegetative states, and fetuses. Next, the class of sentient beings would be even broader, and as we've seen with Singer, might include many non-human animals.[16]

In the next chapter, I offer an expanded version of Realism, where *intrinsic value or moral worth* is the salient feature. By viewing Realism that way, we can understand the Realist presuppositions that lay behind attempts to ground legal rights on moral conceptions. For as we will see, there is a deep connection between our conception of rights and intrinsic value. And so if we were looking for a way to capture what is essential in making something deserving of legal rights, we might go looking for ways to establish that they have moral rights and piggyback our legal conceptions on our moral ones. That, as I will show later, is one of the prominent ways (Realist) RoN advocates argue for their view—by grounding their arguments for legal personhood and rights on *moral* conceptions. I will call that kind of reasoning into question as well.

Legal Rights and Anthropomorphism

Before examining in some detail the anthropomorphism that creeps into Stone's account and others, I want to present additional examples of those who wish for nature to have rights and who also make use of anthropomorphic conceptions in their descriptions of nature. The point here is not yet to critique this language but to simply point out how widespread it is. So we find, for instance, Cullinan, in his defense of Earth jurisprudence, writing of nature's complexity:

> The pale termites were working quickly and *purposefully* to repair the damage and to stop the rapidly warming sunlight streaming in. The workers moved back and forth, efficiently sealing the hole with spittle and earth. The soldier termites positioned themselves at *strategic intervals* so that they could protect the workers and prevent invaders gaining access to the nest. . . . Each termite *knew* exactly what it needed to do, and the actions of each tiny individual combined seamlessly to produce the outcome that the welfare of the community *demanded*.[17]
>
> These terms are not capable of simple definition, but for our purposes we can understand "differentiation" as referring to an inherent tendency

16 Of course, this is where we will find some of the debates about which beings do belong in each category, with such questions arising as whether fetuses and those in vegetative states do, in fact, feel pain and are, therefore, sentient. We can sidestep those debates given the larger issues at stake.

17 Cormac Cullinan, *Wild Law: A Manifesto for Earth Justice* (Bloomsbury Publishing 2011) 28.

towards diversity, variation and complexity; "autopoiesis" as an inherent ability to self-organise and to be *self-aware*.[18]

Others, like the Catholic priest Teilhard de Chardin and Peter Russell, author of The Awakening Earth (1982), see Earth as an organism that is evolving towards being a super-organism. Having made leaps from energy levels to matter, and from matter to life, Earth is now poised to make a leap to a form of ordered global *consciousness*.[19]

Cullinan is certainly not alone.

If the forest is alive—if the animals, the plants, the fungi, the river, the air, and the rocks are all animate beings—then we need to find ways to hear their *voices and spirits*. The rights of nature paradigm included in the Ecuadorian Constitution, which is now incorporated into laws and court rulings around the world, is the Western legal translation of the more fundamental notion that everything is alive, that all beings *speak* in their own ways—and that law, science, and spirituality are not mutually exclusive but rather participants in a growing *conversation* about what human rights mean in the Anthropocene.[20]

Equipped with sensors and artificial intelligence (AI) technologies, scientists like David Gruber eavesdrop on *conversations* among whales, birds, bats, and mole rats in order to decipher and translate their *languages*.[21]

In another example of a Realist strategy, if not outright anthropomorphism, we have the important book *Is the River Alive?* by Robert MacFarlane. In this powerful and moving book, he urges us to reconceive rivers as more than just lifeless mud covered by flowing inanimate waters. Drawing on stories and histories from Indigenous Peoples who have lived and thrived side by side with those rivers, he pushes us to conceive them as living units and, as such, are worthy of rights because they are properly the kinds of things that should be recognized as persons in their vitality and dynamism, a vitality that mirrors our own.

I take the Rights of Nature movement at its best to be a kind of legal "grammar of animacy": that is to say, an attempt to make structures of power align with perceptions of a world which is far more alive than power usually allows. "The law," as the Nyikina Warrwa scholar-activist

18 ibid, 88. My emphasis.
19 ibid, 89.
20 César Rodríguez-Garavito, 'More-Than-Human Rights: Law, Science, and Storytelling Beyond Anthropocentrism' in César Rodríguez-Garavito (ed), *More Than Human Rights* (NYU MOTH Project 2024) 23.
21 ibid, 31.

Anne Poelina puts it, "is being used creatively to train human beings to listen, pay attention to, and learn from, rivers." Recognizing nature's rights is one means of trying to tell a different story about the living world: a very old story, given new expression. A story in which the world is "not a a machine after all," as D. H. Lawrence put it, but "alive and kicking."[22]

For those who, like me, have been largely raised on rationalism, to imagine that a river:

> is alive in a way that exceeds the sum of the lives it contains is difficult, counter-intuitive work. It requires unlearning, a process much harder than learning. We might say that the fate of rivers under rationalism has been to become one-dimensional water. Rivers have been systematically stripped of their spirits and reduced to what Isaac Newton called "inanimate brute matter."[23]

With that lead-up in mind, I turn more fully to Stone's seminal piece, "Should Trees Have Standing?" which posed an early and important challenge, asking us to imagine the (then) nearly unthinkable extension of legal rights to natural objects.[24] Stone points out:

> [T]here will be resistance to giving the thing "rights" until it can be seen and valued for itself; yet, it is hard to see it and value it for itself until we can bring ourselves to give it "rights"—which is almost inevitably going to sound inconceivable to large group of people.[25]

In seeking to extend the domain of rights to include the rest of the natural world, Stone is, as Naffine put it:

> [Urging] us to abandon the idea of a natural bearer of rights and duties . . . and asks us to start elsewhere in our thinking about persons. . . . Such analysis might well suggest the assignment of rights to non-human animals, as well as to rivers and trees.[26]

On Naffine's reading of Stone, he is looking to move past the limited view of having only humans as legal persons or at least of having humans as the

22 Robert MacFarlane, *Is a River Alive?* (Hamish Hamilton Press 2025) 36.

23 ibid, 24.

24 Christopher Stone, *Should Trees Have Standing?: Law, Morality, and the Environment* (3rd edn, Oxford University Press 2010) 3.

25 ibid, 3.

26 N Naffine, 'Legal Personality and the Natural World: On the Persistence of the Human Measure of Value' (2012) 3 *Journal of Human Rights and the Environment* 74.

paradigm and model for what a legal person should look like. That might suggest that Stone is rejecting the Realist view we've been discussing. I, however, think Stone has only partially, at best, rejected the Realist model. For on a closer look at how trees and the like could be granted rights, standing, and personhood, Stone reverts to describing nature in rather human-like terms. Consider again the passage I quoted earlier, where Stone is speaking of the needs and wants of his lawn, arguing that it makes as much sense to apply those terms to his lawn as it does to another human being.

> [N]atural objects *can* communicate their wants (needs) to us, and in ways that are not terribly ambiguous. I am sure I can judge with more certainty and meaningfulness whether and when my lawn wants (needs) water, than the Attorney General can judge whether and when the United States wants (needs) to take an appeal from an adverse judgment by a lower court. The lawn tells me that it wants water by a certain dryness of the blades and soil-immediately obvious to the touch-the appearance of bald spots, yellowing, and a lack of springiness after being walked on.[27]

First, it surely is a mistake to equate *needs* and *wants*. For while the former can be understood as specifying the necessary conditions for a thing's existence and survival—a glacier needs cold and snowfall to exist over the decades—*needs* may also be understood as specifying mental states, say a psychological need for safely. *Wants*, on the other hand, are not ambiguous in that way. *Wants* is a psychological term, specifying a particular kind of mental state. Stone is trading on the ambiguity of "needs" in order to smuggle in what I think is an illegitimate appeal to a psychological state, for lawns simply aren't the kind of things that have minds and mental states.

Second, it is notable that Stone finds it important to describe his lawn as having wants and needs. Arguably, this is because he wants a place from which to argue that natural objects have interests, interests which can be harmed, where interests are typically tied to sentience.[28] If he can establish that nature has interests that can be harmed, he will have demonstrated at least the second of the following three conditions he holds necessary for a thing to have rights, namely:

"[A] thing can institute legal actions at its behest, second, that in determining the granting of legal relief, the court must take *injury to it* into account, and third, that relief must run to the *benefit of it*."[29]

27 Stone, *Should Trees Have Standing*, 4.

28 A classic exposition of the idea that interests are found only in sentient beings can be found in Joel Feinberg, 'The Rights of Animals and Unborn Generations' in William T Blackstone (ed), *Philosophy and Environmental Crisis* (The University of Georgia Press 1974). In Chapter 4, I critically examine the link between *interests* and *sentience*.

29 Stone, *Should Trees Have Standing*, 4.

Importantly, as I see it, Stone's talk of his lawn having needs is intended to rebut the objection that speaking of nature's *wants* is merely metaphor.[30] Stone thinks it is not mere metaphor. Thus, he finds it acceptable to attribute something akin to mental states to his lawn. That strikes me as both a clear example of anthropomorphism and a problematic one at that.

Also noteworthy is Stone's characterization of his lawn communicating with him. Surely, there is a difference between i) the lawn's state of being dry and brown that requires water to live, and which Stone observes, and ii) the lawn is communicating that it needs water to live. Given that the lawn is not capable of forming intentions, not to mention that it has no language with which to speak, such ways of talking are metaphorical at best.

The takeaway point here is that, alongside Stone's advocacy for the expansion of personhood to natural objects, we find him still in thrall (at least to some extent) to a Realist framework. It is both surprising and important to discover the depth of the Realist's hold on personhood, as even for those such as Stone wishing to advocate for nature's rights seem to find themselves reflexively relying on viewing non-human nature as relevantly similar to human beings. There again is the irony of seeking to move past anthropocentrism by relying on anthropomorphism.

To be fair, the mere presence of anthropomorphic language is not necessarily damming. Metaphor is important, as is the imaginative role that is played by making nature a continuation of ourselves and what we already value. There are many paths, after all, towards reckoning with the damage we are doing to our planet; be it through argument or image, prose or picture, there are multiple paths to take, all of which have an important role to play. Just the same, when there is heavy lifting being done by such picturesque language, we should be cognizant of whether it is intended as mere metaphor or not. I've suggested that in at least this instance, Stone's anthropomorphic characterizations of nature is intended as more than metaphor. But I also see it as failing.

More fully, in my inclination to speak of "problematic" or "unacceptable" levels of anthropomorphism, I have in mind two particular objections. First, though it may be useful to speak of such things as garage doors halting their progress because they do not *want* to crush objects in their path, in a deeper sense, it seems unreasonable to take that at face value. Garage doors simply aren't the kinds that have mental states such as wants, needs, or desires. So, too, for basic living things: grass simply isn't a complicated enough organism to have mental states. Whatever our theories are about the nature of minds themselves, we should be wary of attributing mental states to things that, on fuller consideration, simply don't and can't have those states. The second problem arises when we find ourselves unsure whether something is sufficiently complicated as to have, for instance, wants.

30 ibid, 11.

Do salmon want to swim upstream? Does an ant want to protect its territory? My simple suggestion and objection here is that, arguably, mental states come as a package. We often explain human behavior by appealing to the *desires* they have and the *beliefs* they have about how the world works.[31] We explain Jones retrieving aspirin from the medicine cabinet by reference to his *desire* to not be in pain *and* his *belief* that the aspirin reduces pain. Mental states come in clusters. Thus, to attribute one mental state to a being seemingly requires attributing a host of them. And if one is not prepared to attribute a host of mental states, one should be wary in attributing even a single one. Do we want to say that the salmon has *beliefs* about the course of the river, as well as desires? That the ant has beliefs about the structural integrity of its nest? Both surely have instincts, but do they have beliefs and desires, not to mention hopes and fears? And while such rhetorical questions do not settle the issue, I do think they should give us pause. After all, if we can explain a thing's behavior more simply, without having to attribute to it a family of mental states, the simpler explanation seems more promising.

Returning to MacFarlane, we could consider his work to be an extended and powerful metaphor—of ways rivers can remind us of the living things that we already think of as worthy of legal protections. There is surely value in that. The solution to our climate crisis is going to require action on numerous fronts, and the power of images and stories cannot be overstated in helping to shape how the public understands this crisis and what kinds of changes are possible. But in the literal sense, surely, the claim is false: rivers don't have the requisite structure, function, and capacity for development found in truly living things.

Second, and this will be discussed further in later chapters, consider how the claim that rivers are (literally) alive might play out in a legal setting where the question of whether a river has suffered harms arises. Were we to proceed on behalf of the river's damages by speaking of how its vitality and life force has suffered, we would be seemingly imagining jurists charged with the task of determining what the essence of life is and whether rivers have it. My assumption is that most jurists would rather not be forced to settle matters of biology, not to mention metaphysics and religion. The point here is that, were we proceeding on Realist footing and holding that rivers deserve rights and personhood, those debates would seem inevitable. For if the threshold for personhood is being sufficiently like a person, jurists would have to decide whether rivers meet that threshold, a task I again assume they would prefer to avoid. On the other hand, were our notions of personhood severed from all conceptions of human beings—be it their sentience, rationality, intrinsic

31 For a thorough examination of the appropriateness of appealing to beliefs and desires to explain the behavior of complex systems, see Daniel Dennett, *The Intentional Stance* (MIT Press 1987).

value, and even life—we would find ourselves freed to decide on different grounds what is deserving of the designation of legal person.[32] That, I'm suggesting, is what the Legalist strategy offers and which I continue developing in subsequent chapters.

Critiques of Anthropomorphism and Assimilation

Besides pointing out the presence of possibly problematic anthropomorphic language, we should also be aware of the perils of the kinds of assimilationist arguments that are being put forth. For in seeking to cast nature as relevantly similar to humans, in order to find justification for granting it the kind of rights that resemble the rights that humans have, we run the risk present in all assimilationist programs. As Beaudry puts it, referring to the argument strategy I mentioned earlier:

> I make this suggestion on the historical and conceptual bases that some outliers defy assimilation, or that assimilating the portion of their identity that *can* be assimilated would not result in a satisfying integration or would not sufficiently protect their welfare. I mentioned that the "integrationist syllogism" has long been a powerful instrument for civil rights movements. However, assimilationist strategies may not be as fruitful for individuals whose agency, capacities, needs and form of life are significantly different from those of the "normal" group, the "inliers." The assimilationist agenda incorporates or excludes outliers based on a benchmark that it does not modify. The result is a failure to show proper concern for beings who do not quite fit the normal mould, by failing to recognize their morally salient characteristics.[33]

The writings of Ward Churchill provide a good example of the dangers of assimilationist projects that appears in international law, particularly in the context of granting rights of self-determination to Indigenous Peoples. Under the "Declaration of the Rights of Indigenous Peoples" (UNDRIP), Indigenous Peoples are said to have such rights, putting them allegedly on the same playing field as all peoples who have been granted that right under the UN Charter.[34] Good intentions aside, the practical implications of UNDRIP reveal that it is not, in fact, a liberational instrument making progress

32 In Chapter 4, I discuss what those alternative grounds might be.

33 JS Beaudry, 'From Autonomy to Habeas Corpus: Animal Rights Activists Take the Parameters of Legal Personhood to Court' (2016) 1 *Global Journal of Animal Law* 18 <https://journal.fi/gjal/article/download/148714/94849/358299>

34 UN General Assembly, *United Nations Declaration on the Rights of Indigenous Peoples: Resolution/Adopted by the General Assembly*, A/RES/61/295, 2 October 2007 <https://www.refworld.org/legal/resolution/unga/2007/en/49353>

towards equality but rather one that "turns the rhetoric of self-determination to the opposite purpose, that of consecrating in law the very structure of internal colonial domination and exploitation at the hands of state entities from which Indigenous nations have been struggling to free themselves."[35]

On Churchill's view, UNDRIP simply reinforces the power of States to ensure that Indigenous People's rights of self-determination are limited, and crucially, that the power and importance of the State remain unchanged and unquestioned. Rather than liberating Indigenous Peoples, on his view, declarations like UNDRIP simply perpetuate the power hierarchy it was supposed to weaken. For it is the State, again, that becomes the arbiter of sovereignty, with the power to grant or withhold independence and self-determination for Indigenous Peoples. At best, this is ironic, given that at first blush, UNDRIP and related legal instruments were thought to be vehicles for removing that power from States and giving it to—or at least sharing it with—Indigenous Peoples.

The morale here seems to be that while the language of "self-determination" may appear liberating, in practice, the notion that all peoples should be subject to the same standards or viewed as all sharing in the same singular goal as decreed by such bodies as the UN ends up frustrating the very goal it alleged to value. Perhaps, instead, the problem is that, when different degrees of power are in play, the project of treating all peoples as similar results in the further entrenchment *of those power differences.*

An analogous implication for nature is clear: in making nature in our image, we run the risk of forcing it into a box it doesn't fit. While it is true we are part of nature, there is much of nature that is radically unlike humanity. Focusing on perceived likeness runs the risk of both forgetting, and more problematically, alienating that which is different. That is the core of Beaudry's concern. Nature, arguably, should be assessed and valued on its own terms, regardless of whether it is like us or not. Anthropomorphism, where not warranted, focuses our attention on the wrong places. And the flip side threatens as well: for if the standard for consideration of legal protections, such as rights and personhood, depends on a perceived resemblance with humanity, that means that things that fail to meet that standard of resemblance are going to be ignored. Given how radically different parts of nature genuinely are from human nature, those parts, having failed to meet the Realist's sought-after resemblance threshold, will simply be left out in the cold. The bifurcation of nature and humanity, which leaves nature in its subservient role as mere object,

35 Ward Churchill, 'A Travesty of a Mockery of a Sham: Colonialism as "Self-Determination" in the UN Declaration on the Rights of Indigenous Peoples' (2014) 20(3) *Griffith Law Review* 526, 527 <https://doi.org/10.1080/10383441.2011.10854709>

not subject, will continue. That is the lesson from Churchill, applied to the power dynamics involving nature.

Likewise, we might take from Churchill the warning that giving rights to nature as a remedy for how it has historically been excluded and alienated from legal protections carries a great irony if we rely on Realist-assimilationist models. For in seeking to make nature like us, we are possibly just reinforcing the idea that, indeed, humanity is the measure of all things. There's a great irony, in other words, in relying on anthropomorphism as a remedy for anthropocentrism.

As discussed earlier, Anna Grear's writings also provide another reminder of the dangers of assimilationist arguments. When we examine, as we've seen, the history of assignments of legal personhood and the patterns of taking the paradigm of personhood to be the propertied white male, we recognize that this becomes the standard to be measured again. Thus, historically marginalized groups have had a chance to be considered full-fledged persons only to the extent that they resemble the paradigm, which typically has led to their exclusion.[36] The gendering of legal personhood has likewise led to the exclusion of people of color. We might say that this entrenched way of construing personhood has led to the "othering" of those who don't fit the paradigm, where that has led to their alienation and exclusion from legal protections, namely, rights.

As we turn to consider the possibility of nature itself having rights, we should be aware of another deeply entrenched habit, namely that the pattern of considering who or what is deserving of personhood and legal rights redounds to the human form. Our would-be persons have to resemble to a sufficient degree the essential features of legal persons as we conceive them, namely that they be rational, or at least sentient.[37] Unless, that is, one considers the personhood of corporations, with their own history and evolution.

So if a would-be person neither looks like a corporation nor a human, one would expect it to be rejected from the realm of legal personhood. By that score, it is not surprising that nature has historically been excluded. It can only be an object, not subject of law. With that in mind, it is also no surprise that (Realist) advocates of nature's personhood and rights would aim to

36 A Grear, *Redirecting Human Rights: Facing the Challenge of Corporate Legal Humanity* (Palgrave Macmillan 2010) 110.

37 Romain Chuffart has pointed out in conversation that in international law, the standard for being a State according to the Montevideo Convention involves, inter alia, "the capacity to enter into relations with other States," which certainly suggests a view of States as able to communicate and reason and thereby makes use of the same kind of assimilation to the human form as the examples I'm discussing. See Montevideo Convention on the Rights and Duties of States. Signed at Montevideo, 26 December 1933. Entered into Force, 26 December 1934. <https://www.ilsa.org/Jessup/Jessup15/Montevideo%20Convention.pdf>

assimilate nature as much to the human form as possible. Assimilation arguments do make sense as a strategy for nature's rights. The problem, to put it broadly, is that nature just doesn't truly fit the mold of humanity, and assimilationist arguments paper over the radically important ways in which nature differs from the human form, whether genetically or in its lack of rationality and sentience. Difference matters as much as similarity, if not more so.

References

Beaudry JS, 'From Autonomy to Habeas Corpus: Animal Rights Activists Take the Parameters of Legal Personhood to Court' (2016) 1 *Global Journal of Animal Law* 18 <https://journal.fi/gjal/article/download/148714/94849/358299>

Bostock D, *Plato's Theaetetus* (Oxford University Press 1988) 152a.

Churchill W, 'A Travesty of a Mockery of a Sham: Colonialism as "Self-Determination" in the UN Declaration on the Rights of Indigenous Peoples' (2014) 20(3) *Griffith Law Review* 526, 527 <https://doi.org/10.1080/10383441.2011.10854709>

Cullinan C, *Wild Law: A Manifesto for Earth Justice* (Bloomsbury Publishing 2011).

Dennett D, *The Intentional Stance* (MIT Press 1987).

Dennett D, 'Jonathan Bennett's *Rationality*' in Eric Schliesser (ed), *Ten Neglected Classics of Philosophy* (Oxford Academic Press 2016) <https://doi.org/10.1093/acprof:oso/9780199928903.003.0010>

Derham D, 'Theories of Legal Personality' in L Webb (ed), *Legal Personality and Political Pluralism* (Melbourne University Press 1958) 5.

Feinberg F, 'The Rights of Animals and Unborn Generations' in WT Blackstone (ed), *Philosophy and Environmental Crisis* (The University of Georgia Press 1974).

Grear A, *Redirecting Human Rights: Facing the Challenge of Corporate Legal Humanity* (Palgrave Macmillan 2010).

Hill TE, 'The Kantian Conception of Autonomy' in JP Christman (ed), *The Inner Citadel: Essays on Individual Autonomy* (Oxford University Press 1989) 91.

Hume D, *Dialogues Concerning Natural Religion* (Clarendon Press 1935) Part VII.

Khattar J, Calvo P, Vandebroek I, Pandolfi C and Dahdouh-Guebas F, 'Understanding Interdisciplinary Perspectives of Plant Intelligence: Is It a Matter of Science, Language, or Subjectivity?' (2002) 18 *Journal of Ethnobiology and Ethnomedicine* 41 <https://doi.org/10.1186/s13002-022-00539-3> accessed 16 February 2025.

MacFarlane R, *Is a River Alive?* (Hamish Hamilton Press 2025) 36.

Montevideo Convention on the Rights and Duties of States, *Signed at Montevideo*, 26 December 1933. Entered into Force, 26 December 1934 <https://www.ilsa.org/Jessup/Jessup15/Montevideo%20Convention.pdf>

Naffine N, 'Legal Personality and the Natural World: On the Persistence of the Human Measure of Value' (2012) 3 *Journal of Human Rights and the Environment* 74.

Regan T, 'The Case for Animal Rights' (1985) 8(4) *Human Studies* 389.

Rodríguez-Garavito C, 'More-Than-Human Rights: Law, Science, and Storytelling Beyond Anthropocentrism' in C Rodríguez-Garavito (ed), *More Than Human Rights* (NYU MOTH Project 2024).

Stone CD, *Should Trees Have Standing? Law, Morality, and the Environment* (3rd edn, Oxford University Press 2010).

Tam K-P, Lee S-L and Chao MM, 'Saving Mr. Nature: Anthropomorphism Enhances Connectedness to and Protectiveness Toward Nature' (2013) 49(3) *Journal of Experimental Social Psychology* 514 <https://doi.org/10.1016/j.jesp.2013.02.001>

UN General Assembly, *United Nations Declaration on the Rights of Indigenous Peoples: Resolution/Adopted by the General Assembly*, A/RES/61/295, 2 October 2007 <https://www.refworld.org/legal/resolution/unga/2007/en/49353>

Williams M, Whitmarsh L, Mac D and Chríost G, 'The Association Between Anthropomorphism of Nature and Pro-Environmental Variables: A Systematic Review' (2021) 255 *Biological Conservation* 2 <https://doi.org/10.1016/j.biocon.2021.109022> accessed 18 January 2025.

Chapter 3

Persons, Rights, and Values

Legal Personhood Via Moral Considerations

The previous chapter explored and sought to undermine the Realist paradigm for what is salient in analyzing candidates for personhood, arguing that it often relies on problematic anthropomorphism. In this chapter, I argue that the Realist model of what matters in personhood goes even deeper and broader than the elevation of such traits as rationality or sentience. The expanded version of Realism I explore allows us to see another layer in which Realism often pervades arguments for nature's personhood, in this case, the use of value and moral notions. I wish, that is, to critique Realist-based arguments for nature's alleged deserving of personhood via moral or value considerations, considered most broadly.[1]

In particular, I'm interested in arguments which suggest that because nature has *intrinsic value* or *intrinsic worth* that it is deserving of legal personhood. Seeking to ground nature's legal rights in notions of moral rights is a quite common theme and strategy for rights-of-nature advocates, though in this chapter, I will be critical of those efforts. Alfonso Donoso, however, nicely articulates a positive reason for linking the legal and the moral:

> [T]he rights of nature are primarily a moral enterprise. Their recognition entails a recognition of value—the value of nature—that is considered important enough to impose duties on individuals, collectives, and institutions that, accordingly, must constrain their actions and practices as demanded by these rights.[2]

1 The philosophical investigation of values is known as *axiology*. A historically based introduction to that field can be found in Findlay, John Niemeyer, *Axiological Ethics* (St. Martin's Press 1970).

2 Alfonso Donoso, 'Towards a New Framework for Rights of the Biotic Community' in Daniel P Corrigan and Markku Oksanen (eds), *Rights of Nature: A Re-examination* (Routledge 2021) 140, 141.

DOI: 10.4324/9781003740186-4

Before explicitly linking conceptions of *legal* rights and notions of intrinsic value or intrinsic worth, I want to make clear how I see these considerations on behalf of the rights of nature as an extension of the Realist model I presented in Figure 2.1 of the previous chapter. On this expanded view of Realism, what I dub Realism*, it is the *intrinsic value* or *moral worth* that we find first and foremost in humans that is the salient feature that should be captured in our conception of legal personhood. Viewing Realism this way lets us understand the Realist presuppositions that lay behind attempts to ground legal rights on axiological considerations.[3]

Given that there is an intuitive connection between legal and moral rights, to establish the former (and the personhood that might come with it), one might be inclined to find arguments supporting the view that nature has the latter—that nature deserves moral rights. And to establish that, one might want to argue that nature has moral rights because it has intrinsic value. I will suggest, in contrast, that arguments for the intrinsic value of nature are sufficiently problematic as to raise serious questions for attempts to ground legal rights in that manner. To put it simply, there isn't an obviously coherent, plausible way to claim that all of *nature* has intrinsic value. If we can't show that nature has intrinsic value, then we also can't show that nature should have rights because it has that value. If Realism is also implicated in failed attempts to ground nature's rights in intrinsic value, then we have yet another instance of Realism coming up short in seeking to model humanity as the paradigm of legal personhood.

Figure 3.1 illustrates how notions of intrinsic value are connected to our earlier Realist picture of traits deemed essential to humans and thus to legal personhood. Realism* offers an additional feature of humans that some Realists think must be captured in the legal person. On this line of thought, it is our intrinsic value, whether it be displayed in our rationality or our sentience, or simply in our existence as human, which is thought essential and which distinguishes us from the rest of the world. It is this value, says Realism*, that law has traditionally sought to protect and advance. It is our intrinsic value which makes us special and worthy of protection via a scheme of rights. If that is correct, then we have another variant on Realism and another way in which we can assess the merits of putative legal persons, such as nature, asking whether the non-human portions of nature have the requisite intrinsic value to make it worthy of rights and personhood.

3 I will not attempt to rebut *all* efforts to ground legal rights in moral rights and duties. Nor will I seek to undermine the more general project of articulating a new ethical framework that can serve as a foundation for rethinking our moral and legal obligations towards nature, of the sort that has been put forth from such authors as Arne Næss, Aldo Leopold, Thomas Berry, and numerous others. That is a project of its own. Instead, I aim for something much more limited, namely, a critical take on one particular strategy—the use of the weighty concept of *intrinsic value*.

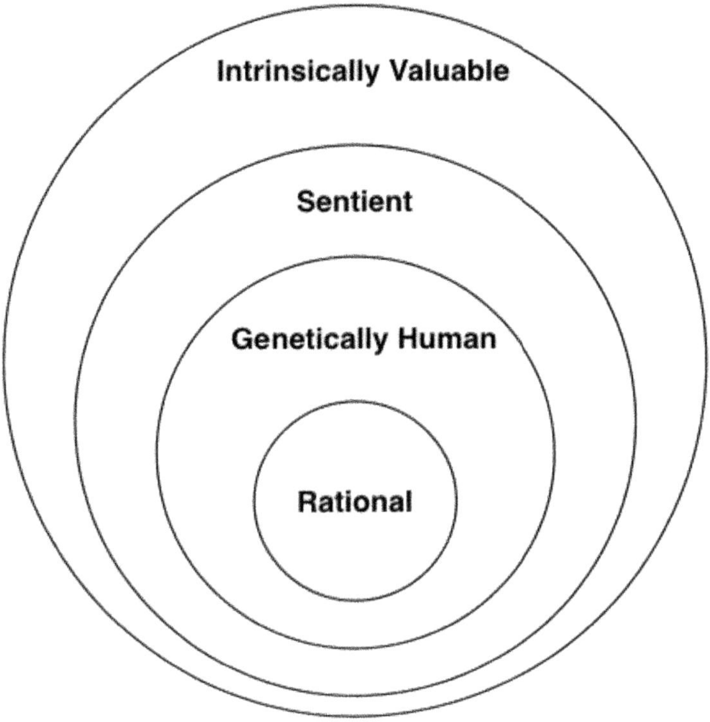

Figure 3.1 Realism*

The link between intrinsic value and being worthy of rights is a deep one. It is also a familiar one if we think though the history of elevating beings not previously considered worthy of rights, of thinking that only things with intrinsic value are deserving of special kinds of treatment, which would include their having rights and being legal persons. Emmenegger and Tschentscher were mentioned in the Introduction, and it is worth remembering their point that "rights and intrinsic value are notions we intuitively connect, attributes we tend to accord jointly or not at all."[4]

Joseph Raz's influential writings on the interest theory of rights is an important example of how rights can be explicitly linked to such evaluative notions as intrinsic value or worth. I mention this in passing now and return

4 Susan Emmenegger and Axel Tschentscher, 'Taking Nature's Right's Seriously: The Long Way to Biocentrism in Environmental Law' (1994) VI(3) *Georgetown International Environmental Law Review* 545, 573.

in detail further later. Raz articulates his understanding of the connection as follows.[5]

> Definition: "X has a right" if and only if X can have rights, and, other things being equal, an aspect of X's well-being (his interest) is a sufficient reason for holding some other person(s) to be under a duty.
>
> Capacity for possessing rights: An individual is capable of having rights if and only if either his *well-being is of ultimate value* or he is an "artificial person" (e.g. a corporation).[6]

The keys here are the criteria for the *capacity* for rights, where Raz explicitly links that capacity not just with having interests, broadly defined, but with beings that have *ultimate value*. That invokes a moral criterion for being a rights-holder. This, in turn, suggests that grappling with the philosophical or moral question of who or what has ultimate value is conceptually prior in debates about who should be considered a rights-holder. Presumably, this is intended to short-circuit objections that the concept of having interests is too broad, allowing one to claim that rare paintings and buildings have interests in not being destroyed and thus have rights. Thus, Raz hopes to exclude them, as they are not the kinds of things that have ultimate value.

Now whether this is a sound move is less pressing at the moment than simply pointing out that this seems to fit squarely within what I'm calling Realism*, that the salient feature for having rights and being a legal person is to have intrinsic value or worth, or in Raz's terminology, to have ultimate value. And as we will see later, while Raz and others hold that nature does not qualify as a rights-holder, they share the Realism* view that the key criterion for being capable of holding rights is that a thing have intrinsic or ultimate value. Raz, then, among many others, see the linkage between rights and value as of particular importance, even if he happens to disagree with those who think nature does have intrinsic value and thus should be accorded rights and personhood. Regardless, the widespread linkage of those two notions makes it particularly important for us to inquire in a more precise manner what exactly intrinsic value involves and how it should be understood. Among the various tasks I think important, which I pursue in subsequent pages, is to:

1) Articulate exactly what is meant by *intrinsic value*
   and
2) Explain which things, if any, do have this privileged status.

5 As we'll see later in this chapter, Raz himself makes a distinction between intrinsic worth and ultimate value, where it is the latter that is doing the heavy lifting.

6 Joseph Raz, *The Morality of Freedom* (Oxford University Press 1988) 166. My emphasis.

What, if anything, belongs in that most general circle in Figure 3.1, and in virtue of what do they belong there?

Now explaining what intrinsic value is, number 1, has been a major challenge for philosophers, environmental ethicists, and legal theorist alike, a challenge that reaches back through the centuries (millennia!) of Western thought. But even should we reach clarity, and even granting that there is a domain of things that are intrinsically valuable, the most pressing challenge for the Realist remains finding whether to include *nature* in that domain—our (2).

As a rough-and-ready start, *contemporary* discussions of intrinsic value are grounded in some of philosophy's oldest distinctions, arguably traceable to Plato's distinction between things that are good in themselves as opposed to things that are good only for their consequences. Commonly, then, a distinction is drawn between what is *instrumentally* valuable and what is *intrinsically* valuable. From here, the discussion immediately becomes difficult, for as a purely philosophical matter, that basic distinction arguably conflates *two* distinct ones.

On one hand, Korsgaard has convincingly argued that the opposite of *intrinsic* value is *extrinsic* value, not instrumental.[7] Extrinsic and intrinsic value point respectively to the *source* of value—where things with extrinsic value derive their value from something more basic, while intrinsic value is, in some ways, foundational or non-derivative.[8] *Instrumental* value, on the other hand, points to a distinction regarding the *manner* in which something is valued. Something with instrumental value is valued merely as a means to an end, as opposed to things that are valued as ends in themselves.[9]

That aside, I suggest that, for the moment, *intrinsic value* is best understood as the value something has in a non-derivative way. It may be said to be valuable for its own sake, independently of whether it is actually valued by others or not. This is intended to capture the sense that some values are objective, not dependent for their worth on whether they are actually valued or prized or desired. *Instrumental value*, in contrast, is a way in which something can have extrinsic, that is, non-intrinsic, value. That is, things with instrumental value derive their value from their being valued as a means

7 Christian Korsgaard, 'Two Distinctions in Goodness' (1983) 92(2) *The Philosophical Review* 69.

8 A dollar bill, for instance, is extrinsically valuable, as receives its value by virtue of it being embedded in a monetary system that assigns different values to differently colored and shaped pieces of paper and bits of metal. It may also be instrumentally valuable, being a means to secure something viewed as more fundamentally valuable.

9 Kant, for instance, famously held that the 'humanity in a person' must always be treated as an end and never merely as a means. Immanuel Kant, *Groundwork of the Metaphysics of Morals* (James W Ellington tr, Hackett Publishing 1993).

towards some end for others. We may value vegetables, for instance, as a means to being healthy, whereas we might value being healthy as an end in itself.

In light of the preceding discussion, the claim that nature has intrinsic value can be understood as holding that it is nature itself that is valuable in a non-derivative way.[10] It has intrinsic value in a way that makes it valuable as something more than a means towards human ends. Statements of nature's intrinsic value are not hard to find amongst writings of environmental ethicists, though, to be sure, they are still subjects of debate. Næss gives a famous straightforward statement:

"The well-being of non-human life on Earth has value in itself. This value is independent of any instrumental usefulness for limited human purposes."[11]

We also find this language in various international environmental legal instruments. Here, we might note the way in which Antarctica is spoken of in the Antarctica Declaration, which seeks to recognize the fundamental, non-derivative value of Antarctica, its intrinsic value, which includes not only the animals, plants, and living things found there but also includes the environment itself.[12]

We also find mention of intrinsic value in the 1992 Convention on Biological Diversity, which asserts in its Preamble that biological diversity has intrinsic value:

"Conscious of the intrinsic value of biological diversity and of the ecological, genetic, social, economic, scientific, educational, cultural, recreational and aesthetic values of biological diversity and its components."[13]

Inclusion of this language in the CBD is significant, as it signals the first time in binding international instruments that biodiversity is recognized

10 As an example of how intricate the debate becomes, John O'Neill has distinguished at least three different meanings of "intrinsic value." See, J O'Neil, 'The Varieties of Intrinsic Value' (1992) 75(2) *The Monist* 119. Bowman, et al. have also distinguished *intrinsic value* from both *instrumental value* and also *inherent value*. M Bowman, P Davies, C Redgwell, *Lyster's International Wildlife Law* (2nd edn, Cambridge University Press 2010) 62. Finally, a fuller discussion of the various positions endorsing the intrinsic value of nature can be found in Benjamin Steyn, 'Nature's Intrinsic Value: A Taxonomy' (2024) 46(1) *Environmental Ethics* 1: 107–130 <https://doi.org/10.5840/enviroethics20245975>

11 A Naess, 'Defence of the Deep Ecology Movement' (1984) 6 *Environmental Ethics* 26. Among other discussions, see L Samuelsson, 'The Cost of Denying Intrinsic Value in Nature' (2022) 44(3) *Environmental Ethics* 267, 268 <https://doi.org/10.5840/enviroethics202292644>

12 "In this Declaration: (a) 'Antarctica' means the community of inter-dependent Antarctic beings that exists South of the Antarctic Convergence, and includes the continent of Antarctica, the ice, sea, seabed, atmosphere, and native species within this area, and the relationships between them." Antarctica Declaration. Draft of 30 November, 2023. Article I, 1(a). <https://antarcticrights.org/wp-content/uploads/2023/11/Antarctica-Declaration_draft_2023.11.30c-1.pdf> accessed 28 May 2024.

13 The Convention on Biological Diversity: United Nations, *Convention on Biological Diversity*, 1992. <https://www.cbd.int/convention/text/>

as having intrinsic value, a value independent of its utility to human interests.[14] Similar language can be found in the Preamble to the so-called Bern Convention:

"Recognising that wild flora and fauna constitute a natural heritage of aesthetic, scientific, cultural, recreational, economic and intrinsic value that needs to be preserved and handed on to future generations."[15]

Finally, we might note the language in the Addis Ababa Principles, as guidelines for implementing the CBD where:

"International, national policies should take into account. . . . Intrinsic and other non-economic values of biological diversity."[16]

Now defining "intrinsic value" as a non-derivative, non-instrumental value as I have done is surely not the end of the story. There is a great literature within moral philosophy, for instance, on whether this is the proper way to capture the essence of this special kind of value.[17] For the sake of argument, let us assume that that meaning suffices. The remaining challenge is to articulate and defend a conception of *nature* which lets us conclude that *it* is intrinsically valuable—valuable in its own right, in a non-derivative way, and not merely as a means to a (human) end or goal.

We saw earlier in our discussions of Singer and Regan that rationality, sentience, and being "the experiencing subject of a life" all have been put forth as grounds for why something has intrinsic value and, thus, as grounds for (some) animals having moral rights (which might, in turn, serve as the foundation for claiming that those animals have *legal* rights as well). In point of fact, though, much of the animal world lacks the traits put forth by Singer or Regan. And unless we are willing to return to those troublesome aspects of anthropomorphism, we will not find those features in the rest of nature: neither in trees, plants, and bacteria, nor certainly in the a-biotic realm of rocks and glaciers. In virtue of what, then, should the bulk of the remaining nature be afforded the lofty notion of intrinsic value?

How do we decide, even in theory, whether something has the kind of value, intrinsic value, that demands recognition and special treatment? If one party to the debate says Antarctica's ice has intrinsic value and should be

14 L Glowka, et al., 'A Guide to the Convention on Biological Diversity' (1994) *IUCN Gland and Cambridge* 9. <https://portals.iucn.org/library/sites/library/files/documents/EPLP-no.030.pdf>

15 Convention on the Conservation of European Wildlife and Natural Habitats * Bern, 19.I X.1979.

16 Conference of the Parties to the Convention on Biological Diversity, Decision VII/12, Annex II: Addis Ababa Principles and Guidelines for the Sustainable Use of Biodiversity, 13 April 2004, UNEP/CBD/COP/DEC/VII/12. Practical principle 10.

17 An extended critique of the notion that biodiversity is intrinsically valuable can be found in JA Newman, GE Varner and SP Linquist, *Defending Biodiversity: Environmental Science and Ethics* (Cambridge University Press 2017).

protected and another disagrees, how are we to settle that debate? One way would be via criteria we use for determining the extension: because x has feature F, it is intrinsically valuable. What then is the relevant criteria? As said, ones offered by Singer and Regan are not broad enough to cover all of nature. So what criteria can allow us to conclude that Antarctica's rocks are intrinsically valuable but a well-designed research station is not? Without guidance here, it seems we are left with a rather unwieldy notion, which was allegedly going to provide the foundation for legal rights.

We should also note some of the skepticism about the very project of grounding environmental policies, practices, and philosophy in that notion of intrinsic value. Some, such as Weston, have argued, in fact, that the search for a univocal account of intrinsic value is either a hopeless pursuit or perhaps, at best, is simply ultimately unhelpful when it comes to solving environmental problems. Perhaps, instead, a more pragmatic approach is to be commended, one that leaves these speculative philosophical inquiries behind.[18]

Raz, Rights, and Values: The Sway of Realism

Having taken a brief tour of some of the complexities, controversies, and challenges for the very notion of intrinsic value, I turn to a fuller examination of Raz's important work on the interlinkage of rights and values. While the preceding discussion has pointed to significant unresolved issues in grounding rights in the notion of intrinsic value, given the challenging nature of the latter, it doesn't pretend to be the end of the matter. Much ink has been spilled by environmental ethicists in seeking to ground rights in moral notions, including ones of intrinsic value or worth. And that will continue. I've merely intended to at least point how difficult the task is. Perhaps the even more important takeaway is that these potential problems, of anthropomorphism and moral groundings, are problems that stem from a Realist perspective on rights and personhood. Put differently, it seems the burdens that comes with Realist premises is quite high. All things being equal, were Legalism not forced to find an answer to these various vexing issues, that would count as a point in the Legalist column. Once Legalism is introduced and defended more fully in later chapters, we'll be in better position to assess the relative merits of each.

To this point in this chapter, we have explored RoN advocates who make use of Realist assumptions or intuitions. I think we can gain a deeper appreciation of the power of Realist inclinations by examining the views of those who use the notion of intrinsic value (and its relatives) *to deny that nature* is the kind of thing that has rights. Here, in particular, I have in mind writings

18 A Weston, 'Beyond Intrinsic Value: Pragmatism in Environmental Ethics' (1985) 7(4) *Environmental Ethics* 321, 316.

of Joseph Raz and Visa Kurki. For, interestingly, the reason they deny that nature is deserving of rights, put simply, is that, on their view, nature lacks intrinsic value. Thus, we find that RoN advocates and also its skeptics are both in thrall to a view which holds that questions of rights and personhood turn to how close the entities in question resemble salient traits of humans. As we've seen, some RoN advocates think nature is sufficiently similar, while others, such as Kurki and Raz, think nature is not. But both sides share the assumption that similarity is the ground on which the issue should be decided. That, again, tells us something important about the hold a Realist view has on legal thinking. That, in turn, shows how radical the Legalist alternative really is.

Let us now delve more deeply, focusing on the influential and much discussed account of interests, rights, and value from Raz, which I began earlier. As mentioned, Raz's preferred terminology is *ultimate value*, for he thinks that there are beings that, while having intrinsic value, do *not* have ultimate value. Yet it is *ultimate value* which is a necessary condition for the capacity of having rights. To bring this out, we need a three-way distinction—between instrumental value, intrinsic value, and finally, ultimate value—and to see how they are connected. By examining their interplay, we will see exactly why Raz denies that things like nature are suitable rights-bearers. Because some of Raz is quite counterintuitive and controversial, I think it important to quote at length.

> Often we ought to or even have duties to act in ways that benefit certain things, and often we ought so to act because of the benefit our action will bring those things. For example, I have a duty to water certain plants because I promised their owners to look after them while they are away on holiday. My gardener has a duty to look after my garden because his contract of employment says so. Some scientists have a duty to preserve certain rare species of plants because they are the only source of a medicine for a rare and fatal disease. In all these cases the people who have duties to act in certain ways have them because it benefits plants. *Yet in none of them is it true that the plants have a right to the benefits. The reason is that in all these cases the benefit is to be conferred on a thing whose existence and prosperity are not of ultimate value. . . .* Having intrinsic value is being valuable even apart from one's instrumental value. But not everything which is intrinsically valuable is also of ultimate value . . .
>
> Consider a man who has a deep attachment to his dog. I share many people's feeling that the relationship is valuable and the man's life as richer and better because of it. Many feel that the relationship is intrinsically valuable. Its value is not just that of a cause of a feeling of security and comfort in the man. Such feelings may be produced by tranquilizers. The relationship is not valued just as a tranquilizer. Its value is in its being a constitutive part of a valuable form of life. Those who share these views

> believe that the existence of the dog is intrinsically valuable. . . . But so far as the story goes the intrinsic value of the dog is not ultimate for it derives from the dog's contribution to the well-being of the man. *The man's well-being is here taken as the ultimate value. The dog non-instrumentally contributes to it. Hence his existence is intrinsically but derivatively valuable.*[19]
>
> My proposed principle of capacity for rights entails that those who regard the existence and well-being of (some) dogs as merely derivatively valuable (even if they believe them to be intrinsically valuable) are committed to the view that dogs can have no rights though we may have duties to protector promote their well-being. For such people dogs have the same moral standing that many ascribe to works of art. Their existence is intrinsically valuable inasmuch as the appreciation of art is intrinsically valuable. *But their value is derivative and not ultimate. It derives from their contribution to the well-being of persons.*[20]

This will strike many, not least Singer and Regan, as simply false. It makes the value of a living being, in this case, a dog, contingent on its contribution to the well-being of its owner. Raz is explicit here that the dog's value is of a piece with the value of a work of art—valuable in virtue of what it offers to a human. What does Raz have to offer to justify such a claim? His answer turns on the justificatory structure of claims about ultimate value. That with ultimate value we can justify claims about things that have intrinsic value, while those very things with ultimate value do not need further explanation or, alternatively, offer justification of their own (ultimate) value.

> The aspects of a good in itself which are of ultimate value are those which explain and justify the judgment that it is good in itself, and which are such that their own value need not be explained or be justified by reference to (their contribution to) other values. The relation of ultimate values to intrinsic values which are not ultimate is an explanatory or justificatory one. Ultimate values are referred to in explaining the value of non-ultimate goods.[21]

Here, it will be useful to draw on the work of Nandi Theunissen, who reconstructs the short argument just quoted and who helpfully brings out the Kantian element behind Raz's claim that it is only human beings that are of ultimate value, as is clear from the dog example. As she reconstructs Raz's argument that ultimate values are the ones that both explain the value of

19 J Raz, *The Morality of Freedom* (Clarendon Press 1986) 177. My emphasis.
20 ibid, 178. My emphasis.
21 ibid, 200.

other things but which themselves do not need to be explained or justified, she draws on a long-used regress argument, where the existence of a self-justifying foundation is needed to prevent an infinite regress of justification or explanation. She writes:

> Here is a statement of the argument in sum: (i) Some things are good or valuable because they can be good for something or someone of value. (ii) There are chains of dependence between things that are valuable in this way. (iii) The chain of dependence between values must come to an end on pain of a vicious regress. (iv) What brings a chain of dependence to an end is something that is such that (a) there are things that are of value because they can be good for it, while (b) its value does not depend on being good for other things—it is good or valuable in itself. (v) Valuers meet conditions (a) and (b). (vi) Therefore, valuers can be the final node in a chain of dependence.[22]

Ultimately, what makes humans—i.e. "valuers"—the final stop of the regress is that humans have the capacity for autonomy. Humans have the capacity for independent judgment, for forging lives of their own choosing that reflects their wishes and values. As autonomous beings, we are able to act as lawgivers; we give to ourselves the principles we will follow as we make decisions about what we should pursue, how we should act, and how we should live. These principles come from us, and that is what makes us free. In contrast, were we to be guided by external forces, we would no long be free, would no longer be autonomous, and instead, would be left with heteronomy, not autonomy.[23] As Theunissen summarizes it: "To offer an intuitive formulation, people are of value because they are beings for whom their life can be an issue."[24]

Raz's account of value puts it squarely within what I've called the Realist* camp, for the bearers of ultimate value are humans, in virtue of their capacity to be valuers, or put otherwise, their capacity for autonomy. Given that, it perhaps cuts no ice against Raz to note that on the theory of interests and value put forth by Singer, Raz is guilty of what Singer calls "speciesism," namely that the interests of humans are to be given priority over other sentient beings simply because they are human. Singer thinks focusing on and elevating one's species is akin to other kinds of discrimination based on sheer matters of biology, be it racism or sexism. Biological differences do not

22 LN Theunissen, *The Value of Humanity* (Oxford University Press 2020) 61.

23 "If the will seeks the law that is to determine it anywhere else than in the fitness of its maxims for its own giving of universal law . . . then heteronomy always results." Kant, *Groundworkm* 4:441.

24 Theunissen, *Value of Humanity*, 86.

signify moral differences or an ordering of interests, he thinks, as much as the racist or sexist might otherwise insist.

As we've seen, Raz takes himself as having defended the view that human's capacity for living autonomous lives is a difference that makes a difference morally and would dismiss the charge of "speciesism" as mere name-calling. On Raz's behalf, we might agree that humans and only humans have this capacity for autonomy. Animals do not. Still, we must ask whether that is a difference that makes a moral (and, ultimately, a legal) difference. Is that difference enough to ground the claim that *only* humans are of ultimate value and that other animals, as Raz claims, owe their value to "their contribution to the well-being of persons"?[25]

Here, we are in danger of a standoff, with Raz saying our autonomy does make a difference in the kind of value we have: Singer and Regan, among others, will disagree. Perhaps we can put pressure on Raz's view by examining more closely the structure of his anti-regress argument. Because Raz is drawing on a familiar and oft-used anti-regress argument, we might benefit by examining the challenges faced by other uses of that kind of argument, such as the cosmological argument, as well as the anti-regress argument Descartes makes in his *Meditations on First Philosophy* in the search for a self-justifying belief that can avoid his skeptical musings.[26] In each case, a chain of dependency is established where unless there is an original starting point, it seems the chain will go on indefinitely. In the cosmological argument, Aquinas argues that everything has a cause, and unless there is a first cause, which is itself self-caused, we face an infinite regress of events and causes. Likewise, Descartes holds that all beliefs must be justified. But unless there is a belief that is self-justifying, we would find ourselves with an infinite regress of beliefs and nothing would ever be justified.

Likewise, Raz is arguing that if everything of value gets its value from something else, we are faced with the problem of there being no ultimate source of value. The only thing that halts this regress, as in the others, is something that is its own source of value. We must find something which is valuable in itself, which can provide the foundation and explanation for all other valuable things. The ultimate source would be our ultimately valuable item, providing the final answer as to why instrumentally valuable things are valuable and what their instrumental value is a means towards. For Raz, the

25 Raz, *Morality of Freedom*, 178.

26 Aquinas's "Second Way" is usually what people have in mind when they speak of "the" cosmological argument, which argues that there must be a first cause, one that is self-causing. *Summa Theologica* (Part I, Question 2, Article 3, Notre Dame: University of Notre Dame Press 1975). Descartes's regress argument, in a search for a self-justifying foundational belief, is found in his *Meditations on First Philosophy* (3rd edn, René Descartes, Donald A Cress tr, Hackett Publishing 1993).

regress-halting move comes by appeal to us, as valuers, human beings, who, as "valuers," have *ultimate value*.

We thus have the image of a chain of dependency, be it causal, or epistemic, or axiological, with that chain ending in a special kind of entity—the self-caused, the self-justified, the self-valuer. In opposition to that image, we might begin by noting that in the domain of *epistemic* justification and the alleged need to find a self-justifying belief, there is a well- established position known as Coherentism.[27] On that view, our beliefs have a weblike structure, with different parts of the web supported and justified by other parts.[28] To explain why a given belief is justified, we turn to other parts of the web, even if, in some cases, our beliefs end up (distantly) relying on each other. Our belief, for instance, that there is a red apple before us is justified by our belief that there *seems* to be one and the more general belief that, in standard viewing conditions, things are often just what they appear to be. As for our belief that standard viewing conditions are reliable, we might justify it by appeal to examples in which our beliefs about how things seem to be have been confirmed by our knowledge of how things are. That, of course, implies that, at times, we do know how things are, though the justification for that kind of belief was what started our quest for justifications in the first place. Put otherwise, what might look to a skeptic as a bit of circular reasoning is, for the Coherentist, really just a point about how, in fact, our justifications of perceptual beliefs work.

Something similar seems to be what Weston has in mind in his ruminations on the alleged need for something intrinsically valuable that is to be the ultimate explanation for the value of other instrumentally valuable items. It would be, that is, to object to premise (iii) earlier that there must be an end to the chain of dependencies. Instead, like the Coherentist in epistemology, the justification, source, and support for valued items may have a holistic, nonlinear structure.

> We need not think of an endless series of means each necessitating the next like a long line of falling dominoes. It is more appropriate to think in quite different terms. Consider a more holistic picture conception according to which values are connected in a weblike way, so that any value can be justified by referring to those "adjacent" to it. On this model there is no ultimate reference or stopping point simply because the series of justifications is ultimately, in a sense, circular: to justify or to explain a value is to reveal its organic place among our others. These justifications need not wind their way only in a single direction or even towards a single type of value.

27 The search for this basic belief that can serve as the foundation for all of knowledge is pursued in Descartes's *First* and *Second Meditations*.

28 Laurence Bonjour, *The Structure of Empirical Knowledge* (Harvard University Press 1985).

> If sometimes I value the mountain air because in it I feel (and am) healthy, other times I value health because it enables me to reach the mountains. If sometimes I value the melancholy glory of the autumn because it mirrors the closure of my own year, other times I value the rhythms of my yearly schedule because they mirror the glories of the seasons.[29]

A second line of objection to Raz's regress argument takes issue with premise (v) on the nature of the items that are deemed "valuable in themselves." Here, we can construct a similarly structured argument but one that yields an entirely different set of self-justifying starting points. With a different starting point, we could then construct an entirely different hierarchy of values but one that moves us away from the anthropocentrism of Raz. Thus, even were we to accept the plausibility of values which are somehow self-justifying, that does not yield a conclusion where those values are ones that put human autonomy at the core. Instead, we can use parity of reasoning to construct a hierarchy that puts something like sentience front and center, which, in turn, would be a way of avoiding Raz's view that animals have only a derivative value.

Consider Mill's theory of utilitarianism. For Mill, the morally right thing to do involves maximizing that which is good.[30] As for what is good, Mill holds that it is happiness: all good things are good either because they are themselves instances of happiness or are the reason or explanation why other good things are good. Thus, since the right action is that which maximizes that which is good, it follows that the right thing to do is to maximize happiness.

> The creed which accepts as the foundation of morals "utility" or the "greatest happiness principle" holds that actions are right in proportion as they tend to promote happiness; wrong as they tend to produce the reverse of happiness. By happiness is intended pleasure and the absence of pain; by unhappiness, pain and the privation of pleasure . . . pleasure and freedom from pain are the only things desirable as ends; and that all desirable things (which are as numerous in the utilitarian as in any other scheme) are desirable either for pleasure inherent in themselves or as means to the promotion of pleasure and the prevention of pain.[31]

Put this way, we can see how this mirrors the structure of the anti-regress argument from Raz. Fortune and fame, if good, are so because they

29 Weston, 'Beyond Intrinsic Value,' 325.

30 For simplicity, I consider only act utilitarianism, though we could just as easily construct the argument that follows along the lines of rule utilitarianism.

31 JS Mill, *Utilitarianism* (Oxford University Press 1861/1998) 66.

are means to happiness. As for the goodness of happiness itself, there, we seem to find the only justification for why happiness is good is simply that happiness is good in itself, which is to say, that the goodness of happiness is self-justifying, and there is nothing further needed to explain what makes happiness good. Whereas Raz gave us human valuers as the regress halter, that which explains or justifies its own value, here, we have happiness as the source of all that is good.

Putting the two regress arguments side by side, we find that this alternative argument has the same structure, in that it is demanding something that halts the regress of justification or explanation of why something is good or valuable but that it yields a very different account of values, interests, and potentially of the foundations of rights. If we decide that happiness itself is what matters, regardless of whose happiness it is, we end up with something like Singer's view, developed in his influential book, *Animal Liberation.*

> The capacity for suffering and enjoyment is a prerequisite for having interests at all, a condition that must be satisfied before we can speak of interests in a meaningful way. It would be nonsense to say that it was not in the interests of a stone to be kicked along the road by a schoolboy. A stone does not have interests because it cannot suffer. Nothing that we can do to it could possibly make any difference to its welfare. The capacity for suffering and enjoyment is, however, not only necessary, but also sufficient for us to say that a being has interests—at an absolute minimum, an interest in not suffering. A mouse, for example, does have an interest in not being kicked along the road, because it will suffer if it is.[32]

The point of this alternative anti-regress argument is that even if we find agreement on what we mean by "intrinsic" or "ultimate" value, we can see that Raz's attempts to ground a theory of rights in the interests of humans involves a choice. It is a choice of *whose interests are to count as foundational and fundamental.* By juxtaposing Raz alongside Mill and Singer, we discover that we are not compelled to carve up the value space as Raz does. Using his own form of reasoning, which involves looking for an ultimate good that explains the goodness or value of everything else, we can just as plausibly generate an account of interests that moves beyond Raz's anthropocentrism. Put somewhat differently, intrinsic and ultimate value seem to be in the eye of the perceiver, and one not already committed to a theory that places human interests and human-like qualities at the center of the worldview may find little to compel them to make that choice. Nothing in Raz's reasoning mandates we accept his anthropocentrism or his theory of value, much less his foundational claims for what can be a rights-bearer.

32 P Singer, *Animal Liberation* (HarperCollins 1975) 8.

If all of this is correct, attempts to ground legally protected interests and rights in such notions of intrinsic or ultimate value may be far less persuasive than they appear. By way of contrast, as noted numerous times, the Legalist view I'm endorsing frees itself from having to make any of these choices about how to construe and construct hierarchies of value from a foundation of intrinsic or ultimate value. The Legalist, as I see it, is perhaps on more stable footing without the onus of grounding rights (and personhood, for that matter) in such unwieldy moral notions.

The Problem of Realism and Kurki's Theory of Personhood

In Chapter 1, I argued that the best way to understand legal personhood, in a way that makes sense of the use of "person" in law, is to see that term as free of metaphysical commitments. I've sought, in other words, to detangle "person" from any Realist commitments, arguing instead that a so-called Legalist position is better suited to capture the wide range of uses of "person." This chapter, alternatively, has sought to examine arguments about how the person *should* be used—with focus on the idea that nature ought to be considered a legal person. Since this project as a whole is devoted to exploring the relative merits of a Legalist view, I've sought to examine and compare arguments for nature's personhood that rely on Realist assumptions. Namely, the idea that if nature is to be a legal person, then it should resemble, in relevant ways, a human being—that is, a person in the non-legal sense. As we've seen, there are various ways to formulate Realism, including the one I introduced in this chapter, that what matters most about humans and which should be reflected in the law is that a person has intrinsic value.

What I've concluded from this examination is that efforts to tie personhood to an essentialist reading of "person" do not stand up to critical examination. One is either forced to embrace unacceptable levels of anthropomorphism, to rely on problematic assimilationist arguments, or, finally, to problematically ground legal rights in unhelpful notions such as intrinsic value.

In closing, I wish to examine again the writings of Visa Kurki, who offers a detailed and sophisticated account of legal personhood with a number of important distinctions relevant to our discussions. And while he does not wish to argue against RoN tout court, he does think that the granting of legal personality requires much more than simply asserting that "person" is an empty slot that can be filled by anything, and that a judicial or legislative stipulation that something is a legal person is truly sufficient for that status. Instead, he argues that substantive accounts of personhood have to be developed and argued for, particularly if they will have to meet the objections he raises.

What's telling in Kurki's account, for our purposes, is that he seems to share many of the Realist assumptions we've explored. While Stone, as we've seen, seeks to give nature rights on grounds that nature is sufficiently human-like to warrant that status, Kurki (and with him, Raz, as we've seen) wishes to *deny* that nature can have rights, on grounds that it *isn't enough* like humans and, in particular, that it is not sentient. The common thread linking Stone, Raz, and Kurki, despite their differing views about RoN itself, is that all three seem to agree that the standard for whether something should have rights is the extent to which it does or does not resemble a salient feature of humans. Realism, in other words, is the standard to employ, whether one ends up believing nature does or does not merit rights of its own. This again speaks to how deeply entrenched the Realist model is in debates about nature's rights. If Realism ends up being a dead-end, or at least seriously problematic, that would suggest a very different model for establishing nature's rights and personhood may be necessary. And that, of course, is what I've stated I hope to do by bringing Legalism into the debate and seeing what potential it holds.

As for Kurki, were we to ask whether a river is suitable for personhood, we need to distinguish between what he calls, on the one hand, a legal platform and, on the other, the object that the platform attaches to. A particular legal platform is essentially the bundle of rights and duties that are nameable and distinguishable from other bundles, where, for instance, a natural person could be said to have one bundle *qua* natural person but another as one-person corporation.[33] Here, the bundle of rights and duties would differ, and be distinct, though they might be attached to the same physical being. The platform *qua* natural person would differ from that of the person *qua* corporation, as, for instance, the right against self-incrimination might be found in the former but not the latter platform.

In asking whether a particular river could be a legal person, Kurki thinks the proper way to approach the question is by asking what kinds of entities legal platforms can be attached to. That is, to what actually existing items does it make sense to ascribe a bundle of rights and duties? To answer that requires noting that ascribing rights and duties is not an all-or-nothing matter. As we've seen in Chapter 1, contrary to what he calls the "Orthodox View," which sees legal personhood as simply equivalent to being a bearer of rights and duties, Kurki thinks that personhood functions as a cluster property, akin to the way in which there are numerous incidents that form the concept of ownership, where having different clusters of rights might jointly be sufficient for something being a legal person, without there being a clear demarcation of what number of rights are jointly sufficient.[34] What matters, according to Kurki, is that we be clear on which particular rights and duties are being ascribed to an entity and, with that in mind, recognize that there are different categories of rights and duties we must keep track of in hopes of deciding whether something in fact is capable of being a legal person.

33 Visa Kurki, *A Theory of Legal Personhood* (Oxford University Press 2019) 96.
34 ibid, 120.

Kurki distinguishes incidents relevant to what he calls "passive legal personhood" as opposed to "active legal personhood." For the former, what is necessary is that the alleged subject have "claim-rights," which, following in the tradition of Hohfeld, amounts to a subject having rights which impose duties upon another. Active personhood, in contrast, would require a subject to have various legal competencies and responsibilities, as we paradigmatically find with fully competent adult humans.[35]

Crucially, according to Kurki, to even be deemed a passive legal person, that entity (the thing we wish to attach the platform of incidents that involve that being's claim-rights against others), must be kind of thing that has interests. In other words, it must be the kind of thing that can be said to undergo harms, such that injury to that thing is an injury to the thing itself, not merely a harm to others who might have a stake in the condition of the thing in question. A society for the preservation of a building, for instance, might have its own interests affected should the building be destroyed, but it is hard to find conceptual room, he thinks, for claiming that the building itself has suffered the kind of legal injury that would make it appropriate to say it has some kind of claim-right against others. And the reason is simple, according to Kurki: buildings aren't the kind of things that have interests. They don't have interests because they are not sentient. Their non-sentience is indicative of another salient point, namely that without sentience, the building doesn't have status as the kind of thing that has intrinsic value or worth. Kurki here echoes Raz's connecting of rights with sentience and intrinsic worth that we just explored.

> The distinction relies on a moral distinction: sentient beings are of ultimate value, and bodies of water are not. There may be a legal person tasked with protecting the river, but the relevant legal platform cannot attach to the river (in the strong sense), because the river cannot hold claims.[36]

Note, however, exactly how Kurki shares a key presupposition of the Realist view. Namely, that the criteria relevant to deciding whether something can be a legal person turns on whether the thing in question resembles a human being in a relevant manner. For Kurki, the relevant matter would be whether that thing is sentient. Or alternatively, it seems, whether that thing has intrinsic or inherent value/worth.[37]

35 ibid, 113.

36 ibid, 141–142.

37 Given my earlier discussions of Realism and the importance of sentience, it is telling that Kurki seems to use *sentience* and having *moral worth* nearly interchangeably. I take this as confirmation of my construal of a variant of Realism, what I called Realism*, which sees the moral notion of intrinsic value as of a piece with Realist assumptions. That Kurki thinks there is a deep connection is another indication that he is working within the Realist

So while Kurki disagrees with the various RoN defenders I've mentioned, on a deeper level, we see their agreement. Namely, they both see the proper ground to decide upon nature's potential as a legal person as turning on questions philosophical, metaphysical, and moral on the extent and manner in which non-human nature resembles the nature of human beings. Whether we focus on nature's alleged intrinsic value or the extent to which plants are sentient, the relevant issue for all parties to the dispute is determining the extent to which nature resembles humanity, or as I've put it earlier, the extent to which one accepts the anthropomorphism of nature.

Now Kurki does seem to share the skepticism I have about being able to make the case that nature has intrinsic value. But I submit my skepticism goes deeper than his. I've made clear so far that my goal is illustrate the limitations of Realism *tout court* and, as such, wish to question the very terrain on which Kurki wishes to engage with standard defenses of RoN—that the resemblance of nature to humans is the important factor with which to approach the personhood of nature. In contrast, I'm laying the foundations for an approach to nature's personhood freed from Realist suppositions and, in particular, from the assumption that nature's personhood is to be decided by reference to its degree of resemblance to human beings. "Personhood without anthropomorphism" might be the calling card for the position I'm seeking to develop and explore and which I begin more fully in the next chapter.

References

Antarctica Declaration, Draft of 30 November, 2023. Article I, 1(a). <https://antarcticrights.org/wp-content/uploads/2023/11/Antarctica-Declaration_draft_2023.11.30c-1.pdf> accessed 28 May 2024.

Aquinas, *Summa Theologica* (Part I, Question 2, Article 3, University of Notre Dame Press 1975).

Bonjour L, *The Structure of Empirical Knowledge* (Harvard University Press 1985).

Bowman M, Davies P, Redgwell C, *Lyster's International Wildlife Law* (2nd edn, Cambridge University Press 2010).

Conference of the Parties to the Convention on Biological Diversity, Decision VII/12, Annex II: Addis Ababa Principles and Guidelines for the Sustainable Use of Biodiversity, 13 April 2004, UNEP/CBD/COP/DEC/VII/12. Practical principle 10.

The Convention on Biological Diversity: United Nations, *Convention on Biological Diversity* 1992 <https://www.cbd.int/convention/text/>

Convention on the Conservation of European Wildlife and Natural Habitats, Bern, 19.IX.1979.

Descartes R, *Meditations on First Philosophy* (3rd edn, DA Cress tr, Hackett Publishing 1993).

tradition as I've described it. Kurki seems to leave the door open for non-humans to have a sufficient value to possibly be rights-bearers, in a way that Raz does not. Witness Raz's apparent hostility to the canine world.

Donoso A, 'Towards a New Framework for Rights of the Biotic Community' in DP Corrigan and M Oksanen (eds), *Rights of Nature: A Re-examination* (Routledge 2021) 140.

Emmenegger S and Tschentscher A, 'Taking Nature's Right's Seriously: The Long Way to Biocentrism in Environmental Law' (1994) VI(3) *Georgetown International Environmental Law Review* 545.

Findlay JN, *Axiological Ethics* (St. Martin's Press 1970).

Glowka L, et al., 'A Guide to the Convention on Biological Diversity' (1994) *IUCN Gland and Cambridge* 9 <https://portals.iucn.org/library/sites/library/files/documents/EPLP-no.030.pdf>

Kant I, *Groundwork of the Metaphysics of Morals* (JW Ellington tr, Hackett Publishing 1993).

Korsgaard C, 'Two Distinctions in Goodness' (1983) 92(2) *The Philosophical Review* 69.

Mill JS, *Utilitarianism* (Oxford University Press 1861/1998).

Naess A, 'Defense of the Deep Ecology Movement' (1984) 6 *Environmental Ethics* 26.

Newman JA, Varner GE and Linquist SP, *Defending Biodiversity: Environmental Science and Ethics* (Cambridge University Press 2017).

O'Neil J, 'The Varieties of Intrinsic Value' (1992) 75(2) *The Monist* 119.

Raz J, *The Morality of Freedom* (Oxford University Press 1988).

Samuelsson L, 'The Cost of Denying Intrinsic Value in Nature' (2022) 44(3) *Environmental Ethics* 267 <https://doi.org/10.5840/enviroethics202292644>

Singer P, *Animal Liberation* (HarperCollins 1975).

Steyn B, 'Nature's Intrinsic Value: A Taxonomy' (2024) 46(1) *Environmental Ethics* 107 <https://doi.org/10.5840/enviroethics20245975>

Theunissen LN, *The Value of Humanity* (Oxford University Press 2020) 61.

Weston A, 'Beyond Intrinsic Value: Pragmatism in Environmental Ethics' (1985) 7(4) *Environmaental Ethics* 321.

Chapter 4

Legalism, Interests, and Nature's Nature

The previous chapters argued that efforts to tie personhood to an essentialist reading of humanity do not stand up to critical examination. One is either forced to embrace problematic anthropomorphism, rely on troubling assimilationist arguments, or, finally, to ground legal rights in worrisome moral notions such as intrinsic value. In this chapter, I begin laying out an alternative, namely, an account of legal personhood stripped of essentialist claims that require our legal categories to map tidily onto philosophical, religious, or moral conceptions of human beings. This is the Legalist understanding of personhood I've introduced at several points.

In particular, I address the thorny issue of *interests*; an account of personhood and rights for nature has to wrestle with how nature's interests might be characterized such that nature can suffer legal harms needing remedy and redress. This has proven to be a stumbling block for many. As Neil Williams has nicely documented, a persistent obstacle for RoN is finding a way to account for nature itself having interests,[1] given that the notion of *interest* is typically grounded in sentience and in beings who can take an interest in their own well-being.[2] Interest theories of rights, for example, bestow rights on beings who are said to have interests, where that is often explicitly limited to sentient beings.[3] Given the non-sentience of ecosystems, rivers, and glaciers, what Williams helpfully calls "environmental entities," it is prima facie difficult for RoN advocates to claim that nature and these environmental entities could be rights-holders.

While the issue does pose a genuine challenge for RoN in general, I will suggest that a Legalist position fares better than what I will suggest is the

1 Neil Williams, 'The "No Interest" Argument and the Rights of Nature' (forthcoming) *Philosophers Imprint* <https://journals.publishing.umich.edu/phimp/article/id/4583/>

2 J Feinberg, 'The Rights of Animals and Unborn Generations' in William T Blackstone (ed), *Philosophy and Environmental Crisis* (The University of Georgia Press 1974).

3 For a defense of an interest theory of rights, see Matthew Kramer, 'Getting Rights Right' in Matthew Kramer (ed), *Rights, Wrongs and Responsibilities* (Palgrave Macmillan 2001) <https://doi.org/10.1057/9780230523630_2>

DOI: 10.4324/9781003740186-5

standard approach. For that standard approach is typically bound to the Realist picture I've sought to critique in previous chapters. To take a theme from the late Wittgenstein, it is, perhaps, a picture of interests as having to involve human-like elements that has held us captive.[4] This, however, is a picture we can free ourselves from. This chapter aims, inter alia, to provide an alternative picture.

In Chapter 1, I argued that personhood is best understood along fairly minimal lines, namely, as simply denoting the bearer of rights and duties and, thus, to be understood as stripped of all associations with traits deemed essential to human beings. Severing the connection between human beings and legal persons is the key, I've suggested, to properly understanding the meaning of "person" in legal contexts. As a matter of language, however, the linkage between *person* and associated notions of *rights* and *duties* goes deeper. For *person* is itself but one legal term among many in a complicated web of legal language, one that implicates not only *rights* and *duties* but also *liabilities*, *interests*, *powers*, *capacities*, entitlements, and *harms*, to name a few.[5] With that in mind, the Legalist project I've begun to sketch would surely be undermined if the rejection of Realist assumptions about "person" were coupled with Realist assumptions in the understanding and meaning of those other terms. It would be, in other words, a hollow victory to claim that the best understanding of "person" is one detached from features of human beings, only to find features of human beings figuring essentially in the meaning and understanding of those other interrelated legal terms.

In the case of *interests*, the Realist picture seemingly suggests that harms or injury to nature, ones which can provide the basis for rights violations, are going to have to resemble injury and harm to *humans*. In contrast, my Legalist account seeks to strip the Realist assumptions/conditions not only from the notion of personhood but from all interrelated concepts, thereby providing a more stable foundation for the notion that nature can have rights of its own. One broad goal of this project, after all, is to examine the *relative* merits of a Legalist view, comparing it with a Realist account. We will see in this chapter that the Legalist view does fare better than the Realist on these matters, and that counts as progress. Not only does the Legalist fare comparatively better than the Realist, I also claim that nature is, indeed, deserving of rights because it *does* have its own interests. Interests, thus, will be a necessary condition of having rights, as the traditional interest theory holds,

4 "A picture held us captive. And we could not get outside it, for it lay in our language and language seemed to repeat it to us inexorably." L Wittgenstein, *Philosophical Investigations* (Basil Blackwell 1953) §115.

5 Hohfeld's famous analysis of rights seeks to untangle the various meanings of "rights" and their connection with other crucial legal terms. WN Hohfeld, 'Some Fundamental Legal Conceptions as Applied in Judicial Reasoning' (1913) 23 *Yale Law Journal* 16.

though I will give a different and easier to meet standard for what interests involve.[6]

The key to understanding the non-sentient-based notion of interests I will defend involves again considering the wide range of legal entities that have rights and interests, paying attention to how many of those entities have rights and interests despite lacking anything like sentience. Examining the non-sentience-based interests of various types of corporations, where that includes not just business endeavors but also States, churches, NGOs, and the like, will provide us with a way of seeing how nature, too, can have interests despite its lack of sentience. This approach echoes that of Chapter 1, where I suggested we get a fuller picture of what "person" means by considering the broad range of entities that law has deemed to be persons.

Rights and Interests

Theories of rights are often presented as a contest between so-called will (or choice) theories and interest theories. Most generally, rights are said to protect something of value to the rights-holder. The two competing theories offer alternative conceptions of what exactly is to be protected. Loosely, we might say that will theories protect the sovereignty of an individual—their ability to make choices that reflect their values and goals. Interest theories, on the other hand, protect the value of those individuals with interests by affording them rights.[7] Having interests appears as a much easier threshold to meet than requiring something have a will or the ability to choose, much less to be a sovereign being. In that way, interest theories offer the possibility of extending rights to a much larger class of being than will theories.

Interest theories are often said to be preferable to will theories in that the latter prima facie denies rights to beings that aren't sufficiently competent to exercise autonomy or that don't have a will of their own. Thus, it is often said that will theories deny rights to infants, humans with severe mental incapacities, or ones in persistent vegetative states. Whether it is accurate to claim that will theories do, indeed, make such denials or that we should, in fact, deny rights to those beings is a matter of ongoing debate. For our purposes, it seems clear that if *nature* is to have rights, it will most likely be on grounds that nature simply has interests, as even that low bar is challenging to meet. As said, because much of nature lacks the sentience so often held to be essential to having interests, it is prima facie difficult to show that the entire

6 Chapter 7 will examine the objection that by lowering the threshold for having rights—by relying on a fairly weak understanding of interests—we are faced with the uncomfortable conclusion that many more things will be candidates for rights than we might initially want.

7 Matthew Kramer, 'Rights Without Trimmings' in Matthew Kramer, Nigel Simmonds and Hillel Steiner (eds), *A Debate Over Rights: Philosophical Enquiries* (Oxford University Press 2000) 7.

animal kingdom even has interests; asking rivers to be rational or capable of exercising a will is untenable.

As for the possibility of *nature* having rights, Joel Feinberg offers a common objection, one which repeats the sentiment we've already seen in Raz and Kurki.

> Consider a very precious "mere thing"—a beautiful natural wilderness, or a complex and ornamental artifact, like the Taj Mahal. Such things ought to be cared for, because they would sink into decay if neglected, depriving some human beings, or perhaps even all human beings, of something of great value. . . . But we are not tempted in these cases to speak of "thing-rights" correlative to custodial duties, because, try as we might, we *cannot think of mere things as possessing interests of their own*. . . . Custodians may protect it for the sake of a nation's pride and art lovers' fancy; but they don't keep it in good repair for "its own sake," or for "its own true welfare," or "well-being." A mere thing, however valuable to others, has no good of its own. *The explanation of that fact, I suspect, consists in the fact that mere things have no conative life: no conscious wishes, desires, and hopes; or urges and impulses; or unconscious drives, aims, and goals; or latent tendencies, direction of growth, and natural fulfillments.*[8]

Feinberg is arguing here that mere physical objects such as buildings cannot have rights because they don't have interests, and the crucial test for their having interests is that they be sentient, or as he puts it, that they have a "conative life." If that standard for rights is correct, then it will, indeed, be very difficult to establish that trees, rocks, glaciers, and ecosystems have rights: they don't have rights, as they don't have interests, and they don't have interests because they are not sentient.

Feinberg, of course, is not alone in this view. In our discussion of Raz, we saw that his version of an interest theory requires, first, an entity be the kind of thing that *can* have interests, which he thinks applies only to a limited range of beings, ones that have ultimate value.[9] Kurki, as we've seen, has similar concerns about extending personhood and rights to nature.

> We can owe duties to adults, children and nonhuman animals, but not, for instance, to rivers. Rather, our duties can *pertain* to rivers. This distinction

8 Feinberg, 'The Rights of Animals,' 49. My emphasis.

9 "Capacity for possessing rights: An individual is capable of having rights if and only if either his well-being is of ultimate value or he is an 'artificial person' (e.g. a corporation)." Raz, *Morality of Freedom*, 166. I find it noteworthy that Raz's account seems unable to provide a general account of interests, instead relying on a rather awkward bifurcation between so-called *natural* persons and *artificial* persons. I've suggested, following Grear and others, that we'd be better served by recognizing all of them as creatures of law, which we might more helpfully call "legal entities."

relies on a moral distinction: sentient beings are of ultimate value and bodies of water are not . . . the river cannot hold claims.[10]

For my part, ideally, we wouldn't have to rely on this moral criterion or the variously associate distinctions about values in the first place. As I've argued in the previous chapter, such notions as *intrinsic worth* or *ultimate value* are far from well-behaved and arguably create more puzzles and problems than they solve. Instead, I suggest we drill down on the notion of *interest* itself, considering the full range of legal entities that are recognized as having interests in a quest to find a more inclusive understanding of what counts as an interest in the first place.

Interests Beyond Sentience

We've seen how tightly some scholars tie interests to sentience. In contrast, I suggest we look more broadly at how law takes seriously the interests of different kinds of legal entities, not allowing ourselves to be trapped by a picture whereby interests tracks only with sentient beings or ones with moral value.

Law recognizes, after all, a whole range of legal entities that are not sentient. Paradigmatically, we have the category of corporations—where that is not limited to profit-seeking business enterprises but instead simply designates a legal entity whose rights and duties are a separate platform not reducible to the rights and duties of its constituent members. This includes States, NGOs, religious organizations, as well as business ventures. They have rights, typically ones that include the right to own property, to enter into contracts and to sue, not to mention, in different contexts, additional ones such as political speech rights, rights against defamation, rights against unlawful searches, etc. They have those rights to protect, most broadly, their *interests*.

It is not controversial, after all, to speak of States' interests in the context of international relations and international law, whether it is their interest in not having their currency counterfeited or in having their leaders immune from various kinds of prosecutions. We do so without awkwardly trying to describe a State as somehow sentient.[11] Nor is it controversial to speak of a business's interest in having a stable political environment, nor of a church's interest in being able to own property, to seek tax deductions, to determine

10 V Kurki, *Theory of Legal Personhood* (Oxford University Press 2019) 141–142.

11 We can find anthropomorphism even here: "In its most extreme form, the real entity theory suggested that group organizations are real living organisms with a body and a will . . . This organicist conception led to metaphysical speculations and anthropomorphist ideas about the corporate entity that were bizarre and unhelpful." Susanna Ripken, *Corporate Personhood* (Cambridge University Press 2019) 37.

what codes of conduct are appropriate for schools that operate under its charter, etc. There is no prima facie reason to view those claims about interests as less grounded and coherent as speaking of sentient beings that have interests in being free from pain or in having a healthy environment.

Were we to articulate a core meaning of "interest" that captures the earlier examples, a first approximation might be:

i) X has an interest in Y
   means
ii) X has a stake in Y being the case for Y is relevant to X being better or worse off.[12]

A business has a stake in the level of its stock prices, as those prices are relevant to the business being better or worse off. I have a stake in not drinking, as the amount of alcohol I consume is relevant to whether I am in more or less pain the morning after. A State has a stake in how it is viewed in the international community, as being in good standing is relevant to the health and welfare of the State.

With (i) and (ii) in mind, we might then go on to explain the nature of *harm* or *injury* by tying them to the thwarting (actively or passively) of an entity's interests. *Legal* harms, as Feinberg explains, would be harms that are not justified given a ranking of interests where ones with high priority would make permissible harms to those with lower priority.

> The interests of different persons are constantly and unavoidably in conflict, so that any legal system determined to "minimize harm" must incorporate judgments of the comparative importance of interests of different kinds so that it can pronounce "unjustified" the invasion of one person's interest of high priority done to protect another person's interest of low priority.
>
> Legal wrongs then will be invasions of interests which violate established priority rankings. Invasions that are justified by the priority rules are not legal wrongs though they might well inflict harm in the non-normative sense of simple setback of interest.[13]

Now to speak of a State having interests in being judged a good ally in the international community or of businesses having interests in political stability, or a church having interests in being able to purchase land, is to invite an objection

12 Cf. "In general, a person has a stake in X (whether X be a company, a career, or some kind of 'issue' of events) when he stands to gain or lose depending on the nature or condition of X." Joel Feinberg, *Harm to Other: The Moral Limits to Criminal Law* (Oxford University Press 1984) 33–34.

13 ibid, 35.

to my claim that such entities can have interests without being sentient. For it may be argued that the interests of *those* entities are reducible to the interests of the constituents that comprise the larger entity, where those constituent members are themselves sentient. Thus, a State's interest in peace and security may be alleged to be reducible to the interest all sentient members of the State have in peace and security. A business's financial interests may be alleged to reduce to the interests of the various stockholders and board members. A church's interest in being able to secure property amounts to the interest individual church members have in having a place to worship and to send their children for school.

In more common terms, one might advance what has been called an "aggregate theory" of corporations, which sees a corporation—something that has its own legal personality—as really just an assembly or collection of individuals where the whole is *not* greater than the parts. On such a view, the alleged interests of the whole is merely a summation of interests of the parts, typically the interests of individual sentient humans.

> Because the corporate person is really just the sum total of all those individuals, it is not possible to distinguish meaningfully between the interests of a corporation and those of its human constituents. The "rights and duties of an incorporated association are in reality the rights and duties of the persons who compose it, and not of an imaginary being."[14]

This view of corporations, of course, has been heavily debated. Part of what makes it complicated is that "corporation" can mean many things, and not all *corporations* are for-profit business. So while it may make sense to speak of some business ventures as nothing more than an aggregate of individual members interests, that view becomes strained as we turn to larger and larger businesses, where the interests of the aggregate are so widespread that it becomes difficult to say whose interests count and whose don't. Limited liability also suggests a problem with the aggregate view, as the point of the "corporate veil" is to *not* distribute or reduce the liabilities of the whole to the constitutive members. But perhaps most importantly, the aggregate view is under increasing strain when we consider non-business corporations.[15]

14 Ripken, *Corporate Personhood*, 31. Consider as well Mitt Romney's much discussed and ridiculed claim that corporations are people. The context makes clear he thinks corporations are persons with interests because corporations simply *are* aggregates of human beings: "Corporations are people, my friend. Of course they are. Everything corporations earn ultimately goes to people. Whose pockets? People's pockets." See the discussion of Romney and related issues in Nikolas Bowie, 'Corporate Personhood v. Corporate Statehood' (2019) 132(7) *Harvard Law Review* 2009 <https://harvardlawreview.org/print/vol-132/corporate-personhood-v-corporate-statehood/>

15 An overview of various positions on the nature of corporations can be found in David Millon, 'Theories of the Corporation' (1990) *Duke Law Journal* 201 <https://scholarship.law.duke.edu/dlj/vol39/iss2/2>

A State, after all, is an entity that spans decades, if not centuries. The members of that State are in constant flux, as are the leaders of that State. It is hard to make the case that individual human interests are the point of securing various State interests when the identity of State, like that of a sports team, transcends the members that happen to temporarily comprise it. Moreover, the various interests a State has on the international stage, be they described by international law or international relations, such as the interest in maintaining its standing in the international community, much less simple survival, seem difficult, at best, to understand in reductionist terms.[16] A State, for instance, arguably has interests in promoting democracy and securing human rights abroad, where the consequences for members of the home State are, perhaps, negligible at best. Even worse, in a large and diverse society, there may be many competing interests of the members of a State, and a State's international policy, done in the interest of the State itself, may be contrary to the interests of even a majority of constituents' individual interests.[17]

As a logical point, if we can simply find cases of corporate (understood broadly) interests that aren't reducible to the interests of sentient constituents of that corporation, then we will have secured a sense in which legal entities have robust rights and interests, without having to find anything like sentience anywhere in the mix. If so, then we can sever the connection between interests and sentience, contrary to what we've seen in Feinberg and others.

As we continue, we should pay particular attention to the *two* components in our broadened understanding of interest as involving something's having a stake, where again:

i) X has an interest in Y
 means
ii) X has a stake in Y being the case for Y is relevant to X being better or worse off.

In particular, we have something, Y, that is relevant to X's condition—where the obtaining (or not) of Y is relevant to whether X stays the same or changes. What matters, though, is not simply that Y can play a role in changes that X undergoes.

For there is also an *evaluative* component, namely that the changes in X, due to Y (or, alternatively, not-Y), make *X better or worse off*. We have, in

16 For a thorough discussion of the many dimensions of States and their interests, see Markus Beham, *State Interest and the Sources of International Law: Doctrine, Morality, and Non-Treaty Law* (Routledge 2018).

17 For a clear discussion of the different conceptions of State interests, see Martin Wight, 'Interests of States' in David Yost (ed), *Foreign Policy and Security Strategy* (Oxford University Press 2023).

other words, a distinction between changes to X on the one hand and change that are better or worse for X. Any hope of securing nature's interests will have to reckon with both. Yet it is this distinction which perhaps is the most important and the most challenging one—to say not just that nature or an environmental entity has changed but that it has changed for either *better or worse*.

Before we wrestle with that complexity, a few additional words are in order about the importance of the non-sentience-based account of interests I've introduced. Recall how I've characterized the Realist position as tying legal notions such as personhood to various features claimed to be essential to human beings, be it rationality, sentience, or even intrinsic value. I've also suggested that a Realist is going to connect those alleged essential traits not only to the notion of personhood but also to related notions—e.g. that interests are connected with an arguably essential trait of humans, namely that we are sentient beings. I've then suggested by way of contrast that the notion of *interests* is, in fact, far broader than what such Realist accounts focus on. We can easily note the numerous examples of legal entities that have interests that *are in no way* connected with sentience.

In keeping with my view that a Legalist understanding of personhood, interests, rights, etc., is preferable to Realist construals, I think it is telling that interests have so often been limited to sentient beings. Only sentient beings have interests, the story goes. Perhaps because there is such a deeply entrenched cleavage in law between so-called "natural persons" on the hand and "juridical (artificial) persons" on the other that it has been tempting for many to simply focus on the former and dismiss other types as just a quirk of legal systems, where the real conceptual work is done by focusing on those "natural persons," which are the real persons of law. But to echo again the sentiment from Grear mentioned earlier—arguably, *all* legal persons are creations of law, and there is no legal person more "natural" than another. A unified account of the capacity for rights, *inter alia*, would surely be preferable to bifurcation. Chasing that possibility is what I hope a Legalist strategy can accomplish.

Thus, to avoid that fragmentation, and to see law as unified across different domains, I've suggested we need an account of interests not bound to a conception that tracks with human-centered traits. Interests beyond anthropocentrism, as it were. We instead need a way to speak of legal entities that have stakes in certain states of affairs obtaining (or not obtaining). As for RoN, I'm suggesting that an account of interests *not* linked to sentience would offer the potential for establishing that nature, indeed, has interests, despite its lack of sentience. With nature thereby seen, somehow, as having interests, we would be in a position to argue further that those interests, in some cases, are important enough to be protected by giving nature its own set of rights. Finally, with the linkage between rights and personhood in place, as discussed in Chapter 1, we have the ingredients necessary to take a further step and give nature legal personhood.[18]

To be clear, pointing out that we can make sense of non-sentient interests as in the case of corporations is far from demonstrating that we have a working notion of interests appropriate to nature. As the next section indicates, there are still difficult problems lurking in hoping to establish exactly how nature's interests are to be understood, particularly in claiming that nature can be made better or worse. Just the same, making sense of how nature can be made better or worse, or alternatively, that nature can be understood as having interests, is a challenge for *all* RoN positions, Realist or Legalist. That, after all, is what keeps Feinberg, Raz, Kurki, and many others from granting rights to nature. Still, I maintain that developing an account of a non-sentience-based account of interest *is* progress, even if it is *in utero*. Were we to develop and defend this account, we'd thereby have a strong positive case for Legalism, beyond the various negative points made against Realism in the previous chapters. That is the goal in the sections and chapters that follow.

The Challenge of Evaluation

I've argued that we can understand "interests" in the broadest sense by seeing them as invoking the stake something has in a given state of affairs and that that state of affairs is relevant to whether a thing is made better or worse off. It is the latter, what we might call the *evaluative* side, which presently poses a challenge for RoN. For to speak of *nature's* interests would be to hold that it is possible for conditions to obtain which make nature *better* or *worse* off. The problem, put bluntly, is that this requires distinguishing between nature simply undergoing changes and nature undergoing changes that are better or worse for *it*.

To make sense of the latter, it is tempting to speak of nature being *healthy*, or of *flourishing*, or of nature being *whole*, *unspoiled*, in *balance*, or even its having *integrity*. Those notions are all evaluative and imply a standard against which we can compare nature at different times. Ideally, goes the thought, we articulate a standard for what nature, in its wholeness or balance, etc., looks like and then examine how it has changed relative to that standard when different conditions obtain. With enough knowledge about the causes of those changes, we might then be in a position to assign blame and pursue appropriate legal recourse or remedies if those changes are for the worse and are significant enough to qualify as genuine harms.

Intuitively, some biological entities *can* be made better or worse off. Pouring gasoline on my plant makes it worse off relative to what a healthy plant

18 Though as we've seen in the Introduction, it need not be that we proceed from rights to personhood, for we might first establish personhood and, from there, distill out the rights of that natural entity, be it corporation or river or ecosystem.

looks like. Richard Routley famously taps into such intuitions in his "last man" thought experiments.[19] Here, we imagine the last human on Earth laying waste to the natural environment, killing species, fouling the seas, and polluting the air, with our intuition supposed to be that something wrong has been done and that nature itself is worse off, for, by hypothesis, there are no humans left where nature's spoilage might be bad for them. Such intuitive reactions are often powerful, hence, the strong visceral responses people have when they hear of oil spills that pollute waters, of poisons being discharged into the atmosphere, and of streams dried up because the water has been damned—even when we leave aside the harms that may befall sentient beings. Nature itself has been harmed, goes the intuition, for nature has been made worse off.

The challenge for us, of course, is to do more than trigger intuitions or emotive responses and articulate clearly how it is that nature itself has been harmed. Against the skeptic, that is, we need to find a way to explain how nature has not only been changed but also changed for better or worse. After all, Routley's thought experiments may be doing nothing more than triggering in us an aesthetic reaction, where we are dismayed that something *we* assign *aesthetic* value to has been changed. In that case, our visceral reaction to imaging the spoiling of pristine nature would be, perhaps, no more than a projection of our own sensibilities and not a fact about the worsening state of nature itself.[20]

One way to rebut these skeptical worries would be via examination of how law has, in fact, dealt with the issue. Here, we can find instances in both international and domestic law, where standards for evaluating the harms to nature *have* been articulated. Such efforts would help us to gain a foothold in articulating not just that nature has changed but also in the all-important evaluative sense I've described.

To accomplish this, we might turn to different evaluative standards that have been articulated, one in particular being the notion of *ecological integrity* with its well-defined metrics, which are then used to measure degrees of degradation from a previously established a baseline condition.[21] As Luuppala notes, given that:

> the IBI (Index of Biological Integrity) has been applied in different contexts (e.g. different ecosystems, resource management, and public policy) in at

19 Richard Routley, 'Is There a Need for a New, an Environmental, Ethic?' (1973) 1 *Proceedings of the XVth World Congress of Philosophy* 205.

20 See Alan Carter, 'Projectivism and the Last Person Argument' (2004) 41(1) *American Philosophical Quarterly* 51 <http://www.jstor.org/stable/20010138>

21 There is a large body of work on ecological integrity and its standards. See, for instance, Rosemary Mwanza, 'Enhancing Accountability for Environmental Damage under International Law: Ecocide as a Legal Fulfilment of Ecological Integrity' (2018) 19(2) *Melbourne*

> least 86 different countries . . . *ecological integrity* seems the most promising core right upon which to base a derivative right to restoration.[22]

To see how such evaluative notions could be put into practice, consider a ruling by Ecuador's Constitutional Court on the Aquepi River Case.[23] The ruling there provides an instructive example of a non-metaphorical conception of a river's health, and that is arguably what we need—a non-metaphorical way of assessing nature's well-being, with an objective standard that can be used to evaluate how nature (i.e. the Aquepi River) has changed.

According to the Court's ruling:

> The [2008] Constitution establishes that nature is a subject of rights and has the right "to have its existence and the maintenance and regeneration of its vital cycles, structure, functions and evolutionary processes fully respected."[24]

> In particular:

> The *life cycle* is violated, in other words, when the subject is not allowed to have its natural structure, the fulfillment of its functions is impeded and its evolutionary process is disrespected. The effect of the violation is that it is not possible to fulfill the purpose and the harmony of the ecosystem and the relationship between human beings and nature is broken. The abuse of a river could generate social or environmental conflicts that disrupt harmony and coexistence.[25]

Journal of International Law 586; Klaus Bosselmann, 'The Ever-Increasing Importance of Ecological Integrity in International and National Law' in Laura Westra, Klaus Bosselmann, Janice Gray and Kathryn Gwiazdon (eds), *Ecological Integrity, Law, and Governance* (Routledge Press 2018) 225; Rakhyun E Kim and Klaus Bosselmann, 'Operationalizing Sustainable Development: Ecological Integrity as a Grundnorm of International Law' (2015) 24(2) *Review of European Community and International Environmental Law* <https://doi.org/10.1111/reel.12109>

22 Linnea Luuppala, 'Rights-Based Restoration' in Daniel P Corrigan and Markku Oksanen (eds), *Rights of Nature: A Re-Examination* (Routledge Press 2021) 121. Another standard that has been invoked is *system resilience,* which can be "analysed in terms of 'animal movements, gene flow, and response to cycles of natural disturbance' or other features like watersheds and air-sheds." Cymie R Payne, 'Defining the "Environment" and the Principle of Environmental Integrity' in Carlsten Stahn, Jens Iverson and Jennifer Easterday (eds), *Environmental Protection and Transitions from Conflict to Peace: Clarifying Norms, Principles and Practices* (Oxford University Press 2017) 40 <https://ssrn.com/abstract=3312914>

23 Ecuador Constitutional Court, Judgment No. 1185-20-JP/21 <https://ecojurisprudence.org/wp-content/uploads/2022/02/Ecuador_Aquepi-river-case_433.pdf>

24 ibid, Section 42.

25 ibid, Section 65.

There are multiple points being made in the Court's ruling. For our purposes, what is most salient is that a violation of the river's rights occurs when its functions are *impeded* and its evolutionary process is *disrespected*, such that the purpose and harmony is *broken*. Understood non-metaphorically, a river is healthy when it maintains its function, when it is able to maintain its integrity which, in turn, helps sustain a diverse ecosystem. That includes being able to flow at a rate sufficiently balanced between flooding and drought, which allows for sufficient levels of oxygen to sustain aquatic life, while simultaneously being sufficiently free from pollution so as to continue those functions. A healthy river, what's more, will also provide water for human needs and interests as well.[26] Thus, in the Aquepi River case, the river's well-being or health is grounded in non-metaphorical, objective measures that reflect sound ecological principles.[27]

Such talk of function, ecological integrity, and purpose are found throughout RoN jurisprudence as well as in recent work on *ecocide*, which gives us another perspective from which we can assess the health of an ecosystem, allowing us to articulate, investigate, and even prosecute particularly grievous harms against nature. Consider, for instance, proposed amendments to the so-called Rome Statute, which established the International Criminal Court (ICC) in 1998.[28] In 2020, the Stop Ecocide Foundation convened an Independent Expert Panel to consider a formal definition of "ecocide" to be put alongside other crimes covered by the ICC. Among other things, it made use of what I've called *evaluative* notions of nature's condition that can be affected by acts of war, if not other causes.

> For the purpose of this Statute, "ecocide" means unlawful or wanton acts committed with knowledge that there is a substantial likelihood of severe and either widespread or long-term *damage* to the environment being caused by those acts.
>
> For the purpose of paragraph 1:
>
> a. "Wanton" means with reckless disregard for *damage* which would be clearly excessive in relation to the social and economic benefits anticipated;
> b. "Severe" means *damage* which involves very serious *adverse changes*, *disruption* or *harm* to any element of the environment, including grave impacts on human life or natural, cultural or economic resources.[29]

26 That human interests are relevant here does not detract from the importance of the Court's ruling that nature itself has had its rights violated.

27 Richard Norris and Martin Thoms, 'What Is River Health?' (1999) 41 *Freshwater Biology* 197 <https://doi.org/10.1046/j.1365-2427.1999.00425.x>

28 Rome Statute of the International Criminal Court (adopted 17 July 1998, entered into force 1 July 2002) 2187 UNTS 3 <https://www.icc-cpi.int/sites/default/files/2024-05/Rome-Statute-eng.pdf>. Proposed revisions can be found at 'Independent Expert Panel for the Legal Definition of Ecocide, Commentary and Core Text,' June 2021 <https://ecocidelaw.com/definition/>

29 ibid, 1. 2. Article 8. My emphasis.

The very notion of ecocide suggests harms against nature of such a degree that they warrant classification as a criminal matter. The very fact that an action can be said to harm nature implies, of course, that nature itself has interests of its own, independent of its utility for human needs, interests that should be protected. Marc Allen Gray puts it nicely. "Being different from humans does not mean being less worthy of respect. All living things are vulnerable and, in the case of fauna, sentient, and therefore deserving of protection."[30]

The prospect of making wanton destruction of the environment an international crime—of enshrining in law the notion of *ecocide*—is an example of how harms to nature can be assessed and so provide the foundation for remedy. At the conceptual level, ecocide can be seen as a natural counterpart to RoN. If nature does indeed have rights, then their violation warrants punishment of some sort.

> In many ways, a law of ecocide is the logical extension of establishing rights of Nature. The rights of Nature concept has typically not been extended to criminal law. The crime of ecocide is but a very small and specific piece of the rights of Nature puzzle, but it extends the principles to the field of criminal law, establishing that severe environmental destruction is a crime. The right to life is not much worth without the crime of murder. In the same way, Nature's right to exist and thrive is not worth much as long as there is a lack of accountability for mass damage and destruction of eco-systems. A crime of ecocide at the International Criminal Court enforces, or is even necessary to truly implement, rights of Nature.[31]

Coupling RoN with the crime of ecocide then offers a way of grounding various kinds of legal remedies for the rights violations. These could range from restorative measures, to reparations or compensation for losses of resources to a community, not to mention the symbolic significance to "recognize what has been lost."[32] The concept of *ecocide*, after all, is intended to call to mind the horrors of *genocide*, with an eye to taking as seriously as possible those harms against nature.

> We consider the campaign to criminalise ecocide to be a potentially valuable endeavour. If the goal of international criminal law is to condemn crimes that shock the conscience of humanity, then there is expressive

30 Mark Allan Gray, 'The International Crime of Ecocide' (1996) 26(2) *Art. 3 California Western International Law Journal* 225.

31 Pella Thiel and Valérie Cabanes, 'Ecocide Law as a Transformative Legal Leverage Point' in Jenny García Ruales, Katarina Hovden, Helen Kopnina, Colin D Robertson, and Hendrik Schoukens (eds), *Rights of Nature in Europe: Encounters and Visions* (Routledge 2024) 317.

32 Rachel Killean and Damien Short, 'A Critical Defense of the Crime of Ecocide' (2025) 10 *Environmental Politics* 15 <https://doi.org/10.1080/09644016.2025.2492441>

> value in including ecocide on that list. Furthermore, the criminalisation of ecocide could have an important deterrent effect. While it is always hard to establish the actual deterrent effect of criminalisation, research in domestic contexts suggests that criminalisation can have a greater deterrent effect on harmful environmental behaviours than administrative and civil frameworks, while there is emerging evidence that the ICC has had some deterrent effect since its creation.[33]

To return to the main theme: irrespective of the promise for harms to nature being codified as crimes of ecocide, the notion of injury to nature seems well-behaved enough to maintain, as I've suggested, that changes to nature can, indeed, be characterized as ones that affect the quality of nature. Thus, I maintain the idea that nature can be made better or worse seems plausible.

That is not to suggest, however, that I consider the matter easily or fully settled. There are still conceptual worries about this evaluative notion and, thus, with the claim that nature has interests that can be harmed. The plausibility of such notions as *function* and *integrity* face their own skeptics. On the skeptic's view, mentioned before, changes in nature, be it from pollution, climate change, or the building of dams, are just changes, neither good nor bad ones. Nature itself is itself not better or worse off.[34] Instead, changes are just changes.

Standards for articulating an ecosystem's *integrity* similarly face multiple challenges. One is simply the challenge of demarcating where one system ends and another begins (both spatially and also temporally).[35] Another notes that because ecosystems are constantly changing, it seems arbitrary to consider one moment in time as indicative of integrity while another moment lacks integrity.[36]

I tackle some of these objections further later. Importantly, for now, it must be said that *if* articulating such standards and evaluative notions of nature's changes are problematic, then they are problematic for *all* RoN positions. As far as I can see, there is nothing more troubling for a Legalist construal of nature's personhood and rights, compared with a Realist account. Any and all RoN positions will have to wrestle with the problem of articulating how non-sentient nature can have interests and, in particular, how nature can be said to be made better or worse off. The earlier remarks offer a sketch of how

33 ibid, 15. In Chapter 5, I examine the expressive value of law and related notions of expressive harm.

34 Though, again, we may react as *if* it is worse off.

35 I examine this issue later.

36 Yasha Rohwer and Emma Marris, 'Ecosystem Integrity Is Neither Real Nor Valuable' (2021) *Conservation Science and Practice* 4 <https://doi.org/10.1111/csp2.411>. A rebuttal can be found in James R Karr, Eric R Larson and Ellen W Chu, 'Ecological Integrity Is Both Real and Valuable' (2022) 4 *Conservation Science and Practice* <https://doi.org/10.1111/csp2.583>

one might proceed. The advantage of Legalism, to my mind, is that it offers a theoretical alternative, and thus advantage, to thinking that interests *per se* are conceptually linked to interests appropriate to human beings.

Which Nature?

Throughout this book, I have spoken in general of nature's personhood and rights. Surely, though, I have to specify *which* and *what* nature I have in mind. If we are to take seriously the idea that nature's interests can be harmed, and therefore, its rights violated, we need an understanding of what it is in particular that demands redress or remedy. Here, we might speak of capital-N Nature to capture the sense that *all* of nature is being invoked. Alternatively, we may speak of small-n nature, indicating a particular river or forest or ecosystem that has rights or personhood. There are pros and cons in either option, and a schism has emerged in RoN scholarship.

The distinction of "nature versus nature" can be seen in the two most famous RoN examples commonly referenced, that of Ecuador on one hand and New Zealand on the other. Ecuador's 2008 Constitution speaks of nature in general: "Nature, or Pachamama, where life is reproduced and occurs, has the right to integral respect for its existence and for the maintenance and regeneration of its life cycles, structure, functions and evolutionary processes."[37] In New Zealand, by contrast, it is particular regions or natural entities that have been given personhood and rights. In particular, the Whanganui River (Te Awa Tupua) and the Te Urewera forest have had their legal personality recognized.[38] Other examples of localized ecosystems or particular rivers can be seen in India's recognition of the legal personhood of the Ganges and Yamuna rivers, as well as the Aquepi River is in Ecuador.[39]

Alongside our two different models for characterizing nature, we also have different legal structures. In the Ecuador case, we have specific legal rights mentioned, whereas in New Zealand, both the river and forest are given legal personality without mention of particular rights. This is the distinction I mentioned in the Introduction of the ways in which something can be *legally visible*. We also find two different models for how nature is to be represented. In Ecuador, anyone can act on behalf of nature, whereas in the New Zealand case, a board is charged with its guardianship. Kauffman provides a helpful

37 Ecuador Constitution, Article 71. Constitución Política de la República del Ecuador, 20 October 2008 <https://www.refworld.org/legal/legislation/natlegbod/2008/es/101916>

38 Te Awa Tupua Whanganui River Settlement 2014 <https://www.whanganui.govt.nz/About-Whanganui/Our-District/Te-Awa-Tupua-Whanganui-River-Settlement>; Te Urewera Act 2014 <https://www.legislation.govt.nz/act/public/2014/0051/latest/dlm6183705.html>

39 For river rights in India, see 'The Ganges and Yamuna Rivers' <https://uni.oslomet.no/riverrights/india/>. Information on the Ecuador Aquepi River Case can be found at <https://ecojurisprudence.org/initiatives/rights-of-aquepi-river-case/>

overview of the differing aspects. Model 1, as he calls it, covers Ecuador, while Model 2 is intended as a description of the New Zealand case and other similar legal regimes that specify particular regions of nature.

> Two distinct models have emerged for how to structure rights-of-nature legal provisions, including guardianship arrangements. . . . In Model 1, rights-bearing nature is defined extremely broadly; all of nature within the legal jurisdiction is recognized as having rights. These laws identify the specific rights held by nature, such as the rights to exist, to maintain the functioning of ecosystem cycles, and to be restored when damaged. Under these laws, any person may stand to speak for nature. . . . Representation, therefore, is voluntary; no person is legally obligated to speak for nature. Consequently, the model tends to address rights of nature reactively, with people seeking to defend nature's rights in court when violations are imminent or after they have occurred.
>
> The second model is illustrated by rights-of-nature legal provisions in New Zealand, Colombia, and India, among other places. In Model 2, particular ecosystems are recognized as legal persons, establishing them as rights-bearing subjects. In most cases, legal provisions do not delineate specific rights other than those held by all legal persons. To protect those rights, not only are specific guardians appointed to speak on behalf of the ecosystem, but they are obliged to do so in both legal and policy arenas.[40]

How should we think about these different ways of charactering nature? On one hand, Emily Jones has argued that the environmental goals driving RoN are better served by not "fragmenting" nature into discrete regions. Jones's concerns stem from a larger project, which seeks to undercut the deeply entrenched distinction between the active agency of humans and the alleged passivity of the physical world, which renders it merely an object to be exploited. Her particular concern over how to characterize nature seems to be that dividing nature into distinct portions protected by different legal regimes runs the risk of a patchwork of (more or less effective) legal protections that fails to address the larger issue of how nature itself, nature in general, should be understood and protected.[41]

On the other hand, we have the view of Mihnea Tănăsescu, who argues it is a mistake to speak of nature in absolute and universal terms. This view is partly driven by, I believe, his skepticism that there is a single, overarching, and

40 Craig Kauffman, 'Guardianship Arrangements in Rights of Nature Legal Provisions' in Anthony Zelle (ed), *Earth Law: Emerging Ecocentric Law—A Guide for Practitioners* (Wolters Kluwer 2020) <http://ebookcentral.proquest.com/lib/uoregon/detail.action?docID=6460079>

41 Emily Jones, 'Posthuman International Law and the Rights of Nature' (2021) 12 *Journal of Human Rights and the Environment* 76, 89.

unifying motivation for RoN in the first place. Instead, RoN is best seem as a series of localized disputes that are often, if not always, about political structures and conflicts over power. If that is true, then environmental protection is not necessarily the ultimate goal of all the RoN initiatives we find worldwide. Without that as the single goal, the need (or even possibility) to characterize all of nature in one, single, abstract paradigm is doomed to fail.[42] What matters then will be how each of the multiplicity of cases play out, in their local political contexts.[43]

Tănăsescu's skepticism over there being a single abstract conception of nature available to us is also driven by his reflections on the interplay between Western-legal concepts, such as rights, and the language, culture, needs, and beliefs of Indigenous Peoples, who are often the prime movers in RoN initiatives. What's more, even when we find cases, such as in Equador where nature does seem to be spoken of in universalist terms, Tănăsescu urges caution:

> "[The] western view," which previously has sought to generalize human rights, is likewise making its presence felt in the Ecuador Constitution, with a similar desire to attach to nature the universalism which it had previously attached to human rights. The possibility of allowing Indigenous ontology to disrupt the very notion of universalism seems, here, foreclosed.[44]

Thus, even when we find universalizing talk, such as in Ecuador's talk of Pachamama, Tănăsescu thinks we should keep in mind that the ontologies of Indigenous Peoples are typically not characterized in such universal, abstract terms. One must be mindful then of the possibility that Ecuador's Constitution has been heavily influenced by those Western tendencies towards universality and, in that way, may be misrepresenting the Indigenous ontologies that inspired that constitution.[45]

To refer back to Chapter 1 and the idea of legal persons individuated by their particular bundle of rights and duties, we can now speak of our present choice as between, say:

$\text{Person}_{\text{Whanganui River}}$
$\text{Person}_{\text{Te Ureweara Forest}}$
or
$\text{Person}_{\text{Nature}}$ (e.g. $\text{Nature}_{\text{Pachamama}}$)

42 Mihnea Tănăsescu, 'The Rights of Nature as Politics' in D Corriga and M Oksanen (eds), *Rights of Nature: A Re-Examination* (Routledge 2021) 81.

43 Mihnea Tănăsescu, 'Rights of Nature, Legal Personality, and Indigenous Philosophies' (2020) 9(3) *Transnational Environmental Law* 429 <https://doi.org/10.1017/S2047102520000217>

44 ibid, 450–451. Reprinted with permission from Cambridge University Press.

45 More on the challenges surrounding the interplay of such widely divergent worldviews can be found M Akchurin, 'Constructing the Rights of Nature: Constitutional Reform, Mobilization, and Environmental Protection in Ecuador' (2015) 40(4) *Law & Social Inquiry* 93 <https://doi.org/10.1111/lsi.12141>

Clearly, one of Tănăsescu's concerns is that he sees a universalized view as in tension with the worldviews of Indigenous Peoples, one that typically focuses on the relationships found between humans and nature in a localized region. As said, attempting to speak of nature in an abstract, all-encompassing manner is reminiscent of problematic aspirations for universal human rights.[46]

The problem with such aspirations is connected to my earlier discussions over so-called "assimilationist arguments." For despite lofty ideals to include all humans in universal declarations of human rights, the de facto distribution of rights has historically favored propertied white males, to the exclusion of women and people of color. Rights to the latter groups are secured only to the extent that similarities with the paradigm rights-holders can be established. In this way, difference is typically erased or deemed problematic.

In the case of nature, a similar issue arises in seeking a universal conception of nature, where the risk arises that we will simply pick one particular conception (or region) of nature and force all of nature to fit that model—and thus engage in another kind of assimilation. Particularly worrisome here is *what* is often held out as the paradigm of nature, namely, nature as pristine wilderness, untouched by humans. The problem with that, perhaps in keeping with Tănăsescu's worry about excluding the worldviews and lives of Indigenous Peoples, is that nature as untouched wilderness would exclude nature as modified and used by human beings, particularly Indigenous Peoples.[47] In fact, nature as untouched wilderness may be no more than a Western-created myth. In contrast, the Indigenous philosophies explored by Tănăsescu see humans and nature as deeply integrated and connected, developing together.[48]

If Tănăsescu is correct, instead of speaking of Nature, capital-N, in absolute, general, abstract terms, we are better served by speaking of particular regions or particular ecosystems, as we find in the New Zealand case. As we saw from Jones, however, that raises problems of its own, for in addition to her worries about the fragmentation of different environmental legal regimes, we also have what might be called the "boundary problems." Here, we face the challenge of individuating things like ecosystems and articulating criteria where we can specify where one ends and another begins. If there

46 A thorough discussion of the many objections international human rights faces can be found in David Kennedy, 'The International Human Rights Movement: Part of the Problem?' (2001) 3 *European Human Rights Law Review* 245.

47 On the myths of nature as untouched wilderness, see William Denevan, 'The Pristine Myth: The Landscape of the Americas in 1492' (1992) 82(3) *Annals of the Association of American Geographers* 369 <https://www.jstor.org/stable/2563351>; William Cronon, 'The Trouble with Wilderness: Or, Getting Back to the Wrong Nature' (1996) 1(1) *Environmental History* 7 <https://www.journals.uchicago.edu/doi/10.2307/3985059>

48 Tănăsescu, 'Rights of Nature, Legal Personality,' 451.

are different regions of nature, different particularized natures, as it were, we need to be able to demarcate them. Jamieson articulates this problem nicely:

> This problem arises on both temporal and spatial dimensions. Grasslands turn to shrubs and small trees, and then to forests. Presumably these are different ecosystems successively inhabiting the same space. What happens on the temporal borders of succession? Do we have a little of one and a little of another? When it comes to space, the problems become even more difficult. It makes sense to say that a little ecosystem has emerged on the north side of the large rock in my garden. But it also makes sense to say that my garden is an ecosystem, and so is the valley in which I live, and so on. All of this is immensely more difficult in the Anthropocene, when all ecosystems are rapidly becoming "novel" ecosystems.[49]

If all of the regions of nature are related and interconnected—streams to rivers to oceans, species that migrate, and weather patterns and pollution that cross national boundaries—how do we specify the boundaries of a particular system which is to be protected by law and given rights? The clearest problem arises in the spatial context—where are we to draw our boundaries and borders? But as Jamieson notes, there is also a temporal problem. Judging that an ecosystem has changed to a sufficient extent to be deemed one that has been harmed requires specifying how much change over time is "natural" and where the threshold is between such natural processes and changes and those that have been adversely affected by, say, human activity. What should our unit of time when evaluating those changes? Do we judge whether an ecosystem has been harmed using a window of one year? Two? A decade? A quarter century? If ecosystems are able to restore themselves, how long do we wait? How much time is enough to judge that a harm has become serious enough, say, for a charge of ecocide? What time frame counts as "long-term"?[50]

Nature's Boundaries

As many have noted, this problem of individuating ecosystems and articulating where the boundary of one meets another is one of the thorniest ongoing challenges in ecology.[51] Perhaps even worse, not only do we have the problem

49 Dale Jamieson, 'The Rights of Nature: Philosophical Challenges and Pragmatic Opportunities' in César Rodríguez-Garavito (ed), *More Than Human Rights: An Ecology of Law, Thought, and Narrative for Earthly Flourishing* (NYU Law 2024) 106.

50 As noted in our discussion of ecocide, "long-term" means damage which is irreversible or which cannot be redressed through natural recovery within a reasonable period of time.

51 "For almost as long as ecology has been a discipline, it has struggled to define what constitutes an ecological community." M Vellend, 'The Community Concept' in David J Gibson (ed), *Oxford Bibliographies: Ecology* (Oxford: Oxford University Press 2012).

of specifying where one ecosystem ends and another begins, we also face the challenge of specifying who exactly are the constituents of a given ecosystem, the "groups" (i.e. communities) of different species which overlap in time and space and which are not thought to form a tidy, separable unit. Instead, they mix, blend, and transform over time. This, in turn, makes it "very difficult or even impossible to draw objective, non-arbitrary boundaries between different communities."[52] This, in turn, has led to widespread skepticism that the demarcation can even be done.[53]

In the face of these daunting challenges, we might aim for progress by noting the resemblance between the issue of individuating ecosystems (both at a single time and over time) and the famous philosophical puzzle of the Ship of Theseus. Of course, the downside of this strategy is that even if we can map the ecological issue onto a well-worn philosophical problem, the ongoing debates about the latter might make one wary of connecting the two. Nevertheless, I think it can help to see that at least part of the puzzle about ecosystem identity stems from a familiar puzzle that arises whenever we wonder about the identity of an entity over time.

The Ship of Theseus puzzle has us imagine a ship in port, gradually having its boards, planks, and the like replaced with new materials such that, at a later time, all of the original parts and materials have been replaced. Is the ship at the later time numerically identical to the ship that was brought into port? If so, in virtue of what is it the same ship, given that all of its materials are different at a later time? If it is *not* one and same ship at the later time, at what point did the original ship cease to exist, and what is it about that temporal point that made for the change?

Philosophers, of course, have long debated this, and the parallels with possible solutions are strikingly similar to how one might go about identifying ecosystems over time. One "solution," of course, is to just declare that there is no objective answer and that we can only arbitrarily stipulate that the ship has either or not remained the same over time. Identity on this view is perhaps a social construct, or perceiver-dependent, but is crucially not something of which there is an objective fact of the matter.[54] This pessimistic view parallels

52 RC Looijen and J van Andel, 'Ecological Communities: Conceptual Problems and Definition' (1999) 2 *Perspectives in Plant Ecology, Evolution, and Systematics* 210, 211.

53 The pessimism increases when we note how many different ways there are of even defining what an ecosystem is. Newman counts at least 12, while reporting that, "In the mid-1990s Michael Palmer and Peter White reviewed 26 ecology textbook definitions." JA Newman, 'Communities and Ecosystems' in J Odenbaugh and S Linquist (eds), *Routledge Handbook on the Philosophy of Ecology* (Routledge Press 2015) 7. See MW Palmer and PS White, 'On the Existence of Ecological Communities' (1994) *Journal of Vegetation Science* 279. A helpful overview of this issue, among others, can be found in Mauricio Guim and Michael A Livermore, 'Where Nature's Rights Go Wrong' (2021) 107(7) *Virginia Law Review* 1347 <https://www.jstor.org/stable/10.2307/27135599>

54 PT Geach, 'Identity' (1967) 21 *Review of Metaphysics* 3.

the despair we found earlier over resolving the question of the boundaries, definition, and essence of ecosystems. Where we to persist, though, seeking an answer, we might draw on one of two strategies that have been discussed for addressing the Ship of Theseus and apply them to the identity of ecosystems. Let us call one of them the "material" solution and the other the "formal" solution.

On the former view, we might claim that ecosystems are best defined, both at a given time and over time, by reference to the species that are present—either ones that are most abundant or perhaps ones that are *fundamental*, in that they are "species whose impact on ecosystem functioning is considerably greater than its abundance . . . keystone species."[55] I call this the "material" solution, for it focuses on what the ecosystem is comprised of and, thus, is analogous to holding that the ship ceases to be the same ship that came into dock once a sufficient amount of its materials have been removed. How much loss is compatible with identity over time—of the ship and of the ecosystem, especially given that the change is continuous? Here, we might find different answers, though the important point is that the identity conditions of an ecosystem turn on what items are present in that system.

On the second strategy—the formal one—identity of the ship over time is tied not to its materials but to what the ship's function or organization of those materials is. Since each plank that is removed is replaced, the ship, at all times, is still a functioning ship; its materials are in the form or organizational structure of a ship, able to do ship-things. The ship then retains its identity despite losing its original pieces. The parallel with ecosystems might then turn on how the elements that make up that ecosystem are organized and the functions they perform. Important candidates for the essential functions of an ecosystem might include its being organized so as to be a dynamic, self-sustaining system capable of certain kinds of energy transfers and/or nutrient cycling.[56] So long as that dynamic system is sustained in its functioning, the ecosystem continues to exist and only ceases to exist once that functioning is sufficiently collapsed. How much collapse is significant enough will again come with different answers.

This brief foray illustrates the complexity of questions about the essence, identity, and conditions for the persistence of an ecosystem and the nature of the boundaries that separate diverse ecosystems. That, in turn, shows the challenge for those of us who wish to speak of particular regions of nature, as opposed to Nature in its entirety. The identity and boundary problems I've just outlined pose difficult issues that range from semantics to metaphysics, not to mention biology and ecology. And though it may be unsatisfying to quickly take a stand at this point, I take shelter in the belief that wrestling with these

55 Newman, 'Communities,' 17.
56 ibid, 19.

boundary problems is part of the ongoing scientific research program that aims at articulating standards and criteria for both the boundaries *and* health of ecosystems, including the evaluative notions of change discussed earlier.

Perhaps this is overly optimistic given all the issues in play, but on balance, I think that if we are going to ground legal protections for the environment in solid scientific evidence about the environment, for the sake of consistency, I think we should place similar confidence in scientific accounts of ecosystem management and methods for individuating ecosystems.[57] It is a matter of balance, to be sure, and there are no easy answers to those boundary problems. For my part, the worries over the assimilationist arguments that arise for universalized notions of nature trump the philosophical-conceptual worries about how to individuate boundaries. All things considered then, I recommend understanding talk of nature's rights as involving small-n nature to indicate a preference for speaking of particular areas, regions, or ecosystems instead of a universal, absolute, or generic conception. This will be the perspective I use in subsequent chapters where we encounter RoN applied to particular regions of nature.

References

Akchurin M, 'Constructing the Rights of Nature: Constitutional Reform, Mobilization, and Environmental Protection in Ecuador' (2015) 40(4) *Law & Social Inquiry* 93 <https://doi.org/10.1111/lsi.12141>

Banks-Leite C and Ewers RM, 'Ecosystem Boundaries' in *eLS* <https://doi.org/10.1002/9780470015902.a0021232>

Beham M, *State Interest and the Sources of International Law: Doctrine, Morality, and Non-Treaty Law* (Routledge 2018).

Bosselmann K, 'The Ever-Increasing Importance of Ecological Integrity in International and National Law' in L Westra, K Bosselmann, J Gray and K Gwiazdon (eds), *Ecological Integrity, Law, and Governance* (Routledge Press 2018) 225.

Bowie N, 'Corporate Personhood v. Corporate Statehood' (2019) 132(7) *Harvard Law Review* 2009 <https://harvardlawreview.org/print/vol-132/corporate-personhood-v-corporate-statehood/>

Carter A, 'Projectivism and the Last Person Argument' (2004) 41(1) *American Philosophical Quarterly* 51 <http://www.jstor.org/stable/20010138>

Cobben D, Ooms W, Roijakkers N and Radziwon A, 'Ecosystem Types: A Systematic Review on Boundaries and Goals' (2022) 142 *Journal of Business Research* 138 <https://www.sciencedirect.com/science/article/pii/S0148296321009607>

57 Sample literature on the boundary problem includes: Cristina Banks-Leite and Robert M Ewers, 'Ecosystem Boundaries' in *eLS* <https://doi.org/10.1002/9780470015902.a0021232>; Dieudonnee Cobben, Ward Ooms, Nadine Roijakkers and Agnieszka Radziwon, 'Ecosystem Types: A Systematic Review on Boundaries and Goals' (2022) 142 *Journal of Business Research* 138 <https://www.sciencedirect.com/science/article/pii/S0148296321009607>; David L Strayer, Mary E Power, William F Fagan, Steward TA Pickett and Jayne Belnap, 'Classification of Ecological Boundaries' (2003) 53(8) *BioScience* 723 <https://doi.org/10.1641/0006-3568(2003)053[0723:ACOEB]2.0.CO;2>

Cronon W, 'The Trouble with Wilderness: Or, Getting Back to the Wrong Nature' (1996) 1(1) *Environmental History* 7 <https://www.journals.uchicago.edu/doi/10.2307/3985059>
Denevan W, 'The Pristine Myth: The Landscape of the Americas in 1492' (1992) 82(3) *Annals of the Association of American Geographers* 369 <https://www.jstor.org/stable/2563351>;
Ecuador Constitution, Article 71. Constitución Política de la República del Ecuador, 20 October 2008 <https://www.refworld.org/legal/legislation/natlegbod/2008/es/101916>
Ecuador Constitutional Court, Judgment No. 1185-20-JP/21 <https://ecojurisprudence.org/wp-content/uploads/2022/02/Ecuador_Aquepi-river-case_433.pdf>
Feinberg J, *Harm to Other: The Moral Limits to Criminal Law* (Oxford University Press 1984) 33–34.
Feinberg J, 'The Rights of Animals and Unborn Generations' in WT Blackstone (ed), *Philosophy and Environmental Crisis* (The University of Georgia Press 1974).
Geach PT, 'Identity' (1967) 21 *Review of Metaphysics* 3.
Gray MAG, 'The International Crime of Ecocide' (1996) 26(2) *Article 3 California Western International Law Journal* 225.
Guim M and Livermore M, 'Where Nature's Rights Go Wrong' (2021) 107(7) *Virginia Law Review* 1347 <https://www.jstor.org/stable/10.2307/27135599>
Hohfeld WN, 'Some Fundamental Legal Conceptions as Applied in Judicial Reasoning' (1913) 23 *Yale Law Journal* 16.
'Independent Expert Panel for the Legal Definition of Ecocide, Commentary and Core Text' June 2021 <https://ecocidelaw.com/definition/>
Jamieson D, 'The Rights of Nature: Philosophical Challenges and Pragmatic Opportunities' in C Rodríguez-Garavito (ed), *More Than Human Rights: An Ecology of Law, Thought, and Narrative for Earthly Flourishing* (NYU Law 2024) 106.
Jones E, 'Posthuman International Law and the Rights of Nature' (2021) 12 *Journal of Human Rights and the Environment* 76.
Karr J, Larson E and Chu EW, 'Ecological Integrity Is Both Real and Valuable' (2022) 4 *Conservation Science and Practice* <https://doi.org/10.1111/csp2.583>
Kauffman C, 'Guardianship Arrangements in Rights of Nature Legal Provisions' in A Zelle (ed), *Earth Law: Emerging Ecocentric Law—A Guide for Practitioners* (Wolters Kluwer 2020) <http://ebookcentral.proquest.com/lib/uoregon/detail.action?docID=6460079>
Kennedy D, 'The International Human Rights Movement: Part of the Problem?' (2001) 3 *European Human Rights Law Review* 245.
Killean R and Short D, 'A Critical Defense of the Crime of Ecocide' (2025) 10 *Environmental Politics* 15 <https://doi.org/10.1080/09644016.2025.2492441>
Kramer M, 'Getting Rights Right' in M Kramer (ed), *Rights, Wrongs and Responsibilities* (Palgrave Macmillan 2001) <https://doi.org/10.1057/9780230523630_2>
Kramer M, 'Rights Without Trimmings' in M Kramer, N Simmonds and H Steiner (eds), *A Debate Over Rights: Philosophical Enquiries* (Oxford University Press 2000) 7.
Kurki V, *Theory of Legal Personhood* (Oxford University Press 2019).
Looijen RC and van Andel J, 'Ecological Communities: Conceptual Problems and Definition' (1999) 2 *Perspectives in Plant Ecology, Evolution and Systematics* 210, 211.
Luuppala L, 'Rights-Based Restoration' in DP Corrigan and M Oksanen (eds), *Rights of Nature: A Re-Examination* (Routledge Press 2021) 121.
Millon D, 'Theories of the Corporation' (1990) *Duke Law Journal* 201 <https://scholarship.law.duke.edu/dlj/vol39/iss2/2>
Mwanza R, 'Enhancing Accountability for Environmental Damage under International Law: Ecocide as a Legal Fulfilment of Ecological Integrity' (2018) 19(2) *Melbourne Journal of International Law* 586.

Newman JA, 'Communities and Ecosystems' in J Odenbaugh and S Linquist (eds), *Routledge Handbook on the Philosophy of Ecology* (Routledge Press 2025) 7.

Norris R and Thoms M, 'What Is River Health?' (1999) 41 *Freshwater Biology* 197 <https://doi.org/10.1046/j.1365-2427.1999.00425.x>

Palmer MW and White PS, 'On the Existence of Ecological Communities' (1994) *Journal of Vegetation Science* 279.

Payne CR, 'Defining the "Environment" and the Principle of Environmental Integrity' in C Stahn, J Iverson and J Easterday (eds), *Environmental Protection and Transitions from Conflict to Peace: Clarifying Norms, Principles and Practices* (Oxford University Press 2017) 40 <https://ssrn.com/abstract=3312914>

Rakhyun EK and Bosselmann K, 'Operationalizing Sustainable Development: Ecological Integrity as a Grundnorm of International Law' (2015) 24(2) *Review of European Community and International Environmental Law* <https://doi.org/10.1111/reel.12109>

Raz J, *The Morality of Freedom* (Clarendon Press 1986).

Ripken SK, *Corporate Personhood* (Cambridge University Press 2019).

Rohwer Y and Marris E, 'Ecosystem Integrity Is Neither Real Nor Valuable' (2021) *Conservation Science and Practice* 4 <https://doi.org/10.1111/csp2.411>

Rome Statute of the International Criminal Court (adopted 17 July 1998, entered into force 1 July 2002) 2187 UNTS 3 <https://www.icc-cpi.int/sites/default/files/2024-05/Rome-Statute-eng.pdf>

Routley R, 'Is There a Need for a New, an Environmental, Ethic?' (1973) 1 *Proceedings of the XVth World Congress of Philosophy* 205.

Strayer M, Power M, Fagan W, Pickett STA and Belnap J, 'Classification of Ecological Boundaries' (2003) 53(8) *BioScience* 723 <https://doi.org/10.1641/0006-3568(2003)053[0723:ACOEB]2.0.CO;2>

Tănăsescu M, 'Rights of Nature, Legal Personality, and Indigenous Philosophies' (2020) 9(3) *Transnational Environmental Law* 429 <https://doi.org/10.1017/S2047102520000217>

Tănăsescu M, 'The Rights of Nature as Politics' in D Corriga and M Oksanen (eds), *Rights of Nature: A Re-Examination* (Routledge 2021).

Te Awa Tupua Whanganui River Settlement 2014 <https://www.whanganui.govt.nz/About-Whanganui/Our-District/Te-Awa-Tupua-Whanganui-River-Settlement>

Te Urewera Act 2014 <https://www.legislation.govt.nz/act/public/2014/0051/latest/dlm6183705.html>

Thiel P and Cabanes V, 'Ecocide Law as a Transformative Legal Leverage Point' in JG Ruales, K Hovden, H Kopnina, CD Robertson and H Schoukens (eds), *Rights of Nature in Europe: Encounters and Visions* (Routledge 2024) 317.

Vellend M, 'The Community Concept' in DJ Gibson (ed), *Oxford Bibliographies: Ecology* (Oxford University Press 2012).

Wight M, 'Interests of States' in D Yost (ed), *Foreign Policy and Security Strategy* (Oxford University Press 2023).

Williams N, 'The "No Interest" Argument and the Rights of Nature' (Forthcoming) *Philosophers Imprint* <https://journals.publishing.umich.edu/phimp/article/id/4583/>

Wittgenstein L, *Philosophical Investigations* (Basil Blackwell 1953).

Chapter 4 Appendix

Legal Personhood and AI

Realist versus Legalist Construals of AI Personhood

In developing an account of personhood befitting nature, I've made heavy use of a distinction between Realist and Legalist construals of personhood. I've also turned to the example of corporations at key points, in order to demonstrate how personhood operates in non-sentient-based contexts, including an understanding of "interests." The concept of personhood is sufficiently broad, though, that we should perhaps widen our net even further by examining another context where personhood has been debated. In recent years, we've seen increasingly widespread uses of AI in various forms. And with our growing reliance on AI, from using it to write and edit reports, to managing schedules, to creating works of art and music, to its use in business, stock trading, and portfolio management, to self-driving cars, and a host of other applications, questions of whether AI should have rights and whether it should be a legal person has become increasingly pressing.

In this section, I will not attempt to definitively answer the question about whether AI can or should be granted personhood. Instead, I wish to examine in a preliminary way how that debate might be understood. Further, by examining the various issues that arise in that debate, we can also touch base with a broader question: Can AI, as another example of a purported non-anthropocentric legal person, help shed light on the nature of personhood in general?

By way of a start, we should note that there are various kinds of AI, ones often not distinguished. What's more, AI technology is developing at a rapid pace. On the one hand, that can make speaking to its possibilities difficult, as there are regularly updates and revisions to what we even thought possible with AI.[1] On the other hand, having a general framework for thinking about AI along both economic, social, political, moral, and legal lines is

1 For instance, the seemingly sudden emergence of ChatGPT in November 2022 was, for many, a shocking introduction to how advanced AI systems had become.

arguably made more important by the fact that the technology is developing so quickly.

Perhaps the most important distinction to make at the most general level is between so-called "weak AI" and "strong AI."[2] The former typically includes systems that focus on narrow or specific tasks; hence, it also being referred to as "narrow AI." Large language models (LLM) of the kind used in ChatGPT, AI-assisted Google searches, self-driving cars, and facial recognition systems are the most common and familiar examples of weak AI. In fact, the AI systems that presently exist are all weak AI, as strong AI is at, this point, solely theoretical.

Strong AI systems, sometimes spoken of as "artificial general intelligence," would be ones capable of the "mechanical replication of all intellectual functions that are available to humans" but, as of yet, are to be found only in science fiction stories and theoretical discussions of what is possible for AI.[3] Thus, as we pursue the question of the possibility of AI's personhood, we must be clear whether we have in mind the kinds of weak AI systems already in place and which, no doubt, will get more sophisticated, or whether we are speaking of something radically different and wondering about the legal standing of AI systems that do not yet exist. At the time of writing, no major jurisdiction has granted legal personhood to any AI system, and one should be careful in discussing the issue to not let the future possibility of strong AI muddy a discussion about what the legal standing should be of AI in the present weak form it currently exists.[4]

In Chapter 3, I argued that both proponents and also critics of extending personhood to nature can often be seen as relying on a Realist framework. As we saw, while one might argue that nature cannot be a person because it is *not* intrinsically valuable, another may argue that nature *is* intrinsically valuable and, thus, should be considered a legal person. Both positions here are using a feature deemed essential to humanity—in this case, its intrinsic value—as the standard or threshold to be met in assessing candidates for personhood.

Arguably, we can find a similar dynamic at work in debates over whether AI can or should be granted legal personhood. Once again, we find both advocates and critics of that position looking to see the extent to which AI

2 P Bory, S Natale and C Katzenbach, 'Strong and Weak AI Narratives: An Analytical Framework' (2025) 40 *AI & Society* 2107 <https://doi.org/10.1007/s00146-024-02087-8>

3 ibid, 2107.

4 "Despite differing political systems and legal traditions, all major jurisdictions have thus far rejected proposals to endow AI with formal legal personhood. This consensus underscores a shared regulatory commitment to preserving human-centred frameworks of accountability." Joffrey Baeyaert, 'Beyond Personhood: The Evolution of Legal Personhood and Its Implications for AI Recognition' (2025) 2025 *Technology and Regulation* 355, 374 <https://doi.org/10.71265/ssvg8a97>

resembles essential features of humans. While my preferred terminology speaks of Realist and Legalist frameworks, that distinction maps fairly tidily onto one made by Novelli et al., who prefer to speak of *intrinsic* versus *instrumental* rationales for extending personhood to AI.[5] The two categories here are exhaustive and mutually exclusive, with the former looking to the features an AI system has, and whether those features resemble ones that, when found in humans, are deemed sufficient for personhood. Discovery of such features, such as rationality, might give us reason to treat the AI system analogously to how Kant thinks we should treat the humanity in persons, namely, as "ends in themselves."[6]

On the other end of the spectrum, we find justifications for AI personhood based not on such intrinsic features or qualities but based instead on the ends those AI systems can serve. "Here, legal personhood for AI is seen as a tool to support human goals rather than as recognition of any inherent value in the AI itself," where those goals might range from the economic to the political, social, or even aesthetic.[7]

For my part, I consider the "intrinsic" rationale for granting personhood to AI as nearly synonymous with the Realist strategy I've been exploring throughout this book, whether the supposed intrinsic quality in question is the (alleged) rationality of AI, or perhaps even its (alleged) sentience. Both variations are looking to capture something thought essential to human beings and claiming that the presence of *it* in an AI system would be grounds for granting that system personhood.[8]

This strategy for giving personhood to AI echoes the kind of resemblance or assimilationist gambit we saw in arguments on behalf of animals deserving an elevated status from that of mere objects or property: from Singer's claims that animals are sentient to Regan's view where, as "subjects of a life," animals are deemed of sufficient value to warrant bestowal of rights. The details and domains are different, to be sure, but the argumentative strategy is the same: extend the realm of legal persons by recognizing the similarity of a contested category of beings with beings that already have that status. We even find arguments that replicate Singer's views about the history of expanding the moral community, discussed earlier, where perhaps a future step would include AI systems and robots, based on "whatever feature we humans share

5 We'll see later that the overlap between Legalism and instrumentalism in this case isn't exact, especially once we recall that Legalism comes in different versions.

6 Claudio Novelli, Luciano Floridi, Giovanni Sartor and Gunther Teubner, 'AI as Legal Persons: Past, Patterns, and Prospects' (2024) 6 *Journal of Law and Society* <https://doi.org/10.71265/ssvg8a97>

7 ibid, 6.

8 For a defense of the view that AI systems *are* rational, intelligent, and autonomous, see S Yanisky-Ravid and XJ Liu, 'When Artificial Intelligence Systems Produce Inventions: The 3A Era and an Alternative Model for Patent Law' (2018) 39(6) *Cardozo Law Review* 2215.

with them we believe is a basis for ascribing rights or for taking their interest into account."[9]

Likewise, we find rejections of personhood for reasons that replicate the rejection of personhood for animals and ecosystems. We saw in our discussions of Raz and Kurki that we might agree that a particular intrinsic feature *is* the salient criterion but go on to argue that various entities, such as animals or ecosystems, lack that feature and so are not candidates for personhood. In those examples, it was sentence and ultimate value that were key. In the example of robots and AI, a likely criterion is rationality. Here, someone inclined against granting personhood to AI might argue that it is precisely because AI systems (particularly the weak AI ones that presently exist) *lack* rationality that they cannot be legal persons. More generally, we might rely on what Solum has called the "missing something argument," where any of a number of selected criteria viewed as sufficient for personhood are simply found lacking in AI systems—be it having a soul, or consciousness, desires, etc.[10]

The earlier paragraphs demonstrate one strategy for arguing on behalf of AI's personhood, and with it as well, a strategy for arguing against. Both take as their point of departure the need to identify a feature of AI that is relevantly similar to a trait deemed essential to humans. That, as I've said, is a Realist framework for debating the possibility of granting personhood to AI. As we've seen, however, a Realist framework is not the only possibility. I've characterized the alternative as a Legalist one that doesn't require resemblance between humans and putative new legal persons. To develop this framework, I think it helpful to refer back to the distinction noted earlier between "intrinsic and instrumental" rationales for granting personhood to AI, for the instrumental rationale can be seen as another way of articulating what the Legalist cares about: not intrinsic features of the AI per se but, like corporations and other non-humans that have legal personhood, the point of extending personhood to those entities in the first place. Here, we can find deep analogies between reasons that have been used to grant personhood to States, corporations, LLCs, NGOs, and the like. Namely, giving personhood to those entities serves economic-social-political purposes. That is, it is an instrumental justification for granting them personhood.

> In these settings, the designation "legal person" is the conclusion of particular legal arguments, not a rhetorical reason for those legal conclusions. Here the law decides to treat nonhuman entities analogously to human persons in certain ways and circumstances, for example, as parties in a

9 Mark Coeckelbergh, 'Robot Rights? Towards a Social-Relational Justification of Moral Considerations' (2010) 12 *Ethics and Information Technology* 209, 211.

10 Lawrence Solum, 'Legal Personhood for Artificial Intelligences' (1992) 70 *North Carolina Law Review* 12311262 <https://scholarship.law.unc.edu/nclr/vol70/iss4/4>

> lawsuit, or as possessing the juridical ability to assert various rights and powers. . . . In the legal personhood of nonhumans is found the strongest argument that to ascribe legal personhood to an entity is to do no more than to make arrangements that facilitate a particular set of social, economic, and legal relationships.[11]

On the instrumental rationale, the case for making anything a legal person turns on its ability to best serve social interests, understood most broadly. What then might be these extrinsic or instrumental reasons for giving personhood to AI? Here, we find multiple suggestions. For instance, a need for consistency in how contracts using artificial agents are treated.

> [G]iven the great importance of e-commerce in today's business world, and the increasing number of agent-to-agent contracting transactions that take place on a daily basis, such a move would greatly facilitate a standardized understanding of these interactions. Other motivations could arise from the increasing delegation of responsibility for automated decision-making in the administrative law context. Here administrative decision-making powers could be coherently delegated to the artificial agents of the administrative agency.[12]

A particularly interesting debate about AI and personhood arises in the context of debates about intellectual property. In addition to producing text, AI systems have become extremely sophisticated in producing works of art. Given the huge investments involved in building those AI systems, and the growing demand for AI works of art, reflected in some of the prices those works have fetched at auction, one is driven to wonder whether those artworks are copyright protected and whether they should be. If so, who would be the author of those works? Who would be owner of those works? For if those works are copyright protected, the owner would have claim-rights against those who illegally copy or distribute their works. And because I've repeatedly linked being the bearer of rights to being a legal person, we might find ourselves wondering whether a route to legal personhood for AI might come through its right to own the works it (seems to) create.

AI, Art, and Copyrights

If an AI system creates a work of art, should there be copyright protection for that work? In general, copyright protections serve the social-economic

11 Samir Chopra and Laurence F White, *A Legal Theory for Autonomous Artificial Agents* (University of Michigan Press 2011) 158.

12 ibid, 161.

function of encouraging innovation and of protecting the "fruits of one's labor" of the creator, while also benefitting society at large that is able to enjoy and make use of those individuals' creative efforts.

Assuming that it is appropriate to speak of "creation" and "artwork" in this context, we face a myriad of questions, many of which have implications not just for the local issue of this or that copyright protection but also the larger issue of potential personhood for these "creative" systems. For instance, if an AI system is deemed to have created a work of art, would it make sense to give the AI system a right to its creation? Or should it go to the developer of the AI system? Or perhaps to the end user who used a series of prompts with the AI system to generate the artwork? If one shudders at all of these options, would we be content to simply conclude that there really is no author or creator and that the work should immediately go into the public domain?

Here is where the instrumental argument becomes important: continuing to deny that AI systems are genuine authors might undermine the financial incentives for creating AI systems. Were that to happen, it would presumably affect not only the development of AI systems that can create art but also a host of related AI systems, such as ones that can map and model the intricate folding patterns of proteins and that have been used to create breakthroughs in biology and medicine.[13] Thus, if nothing else, the huge amounts of money at stake in AI art provides an instrumental reason to sort out who, if anyone or anything, should be considered the author and owner of those works since, among other things, huge investments in the innovation of AI are at stake.[14]

As Guadamuz has helpfully shown, across different jurisdictions, we find an array of solutions to the authorship issue, with some such as the United Kingdom, providing statutory protections of computer or AI-generated works.[15] The United States, on the other hand, seems to be an outlier, as courts have repeatedly affirmed the view that only human beings can be the author of works and, thus, potential copyright holders. In a widely noted recent case, *Thaler v. Vidal* (2022), the US Federal Circuit explicitly reaffirmed the view that *only* humans meet the requirement for legal authorship and denied a copyright of a painting, *A Recent Entrance to Paradise*, that Dr. Stephen Thaler claimed was produced solely by an AI machine of his own creation, the "Creativity Machine."

13 J Jumper, R Evans, A Pritzel, et al., 'Highly Accurate Protein Structure Prediction with AlphaFold' (2021) 596 *Nature* 583–589 <https://doi.org/10.1038/s41586-021-03819-2>

14 The AI art market was valued at approximately $3.2 billion in 2024. <https://garagefarm.net/blog/ai-is-art-doomed>

15 Andres Guadamuz, 'Do Androids Dream of Electric Copyright? Comparative Analysis of Originality in Artificial Intelligence Generated Works' in Jyh-An Lee, Reto M Hilty and Kung-Chung Liu (eds), *Artificial Intelligence and Copyright Law* (Oxford Academic Press 2021) <https://doi.org/10.1093/oso/9780198870944.003.0008>

On the surface, the denial is straightforward, with the Court claiming that Creativity Machine "cannot be the recognized author of a copyrighted work because the Copyright Act of 1976 requires all eligible work to be authored in the first instance by a human being."[16] A closer look at the opinion shows the Court venturing into some of the more philosophical debates it is often loath to engage in. The reasoning notes, first, that machines in general lack intentions and a wide range of other features that are typically found in humans. Second, the absence of those features in machines, such as the Creativity Machine, entails that it cannot properly be viewed as an author.[17] Having read to this point, it will not surprise the reader that I find this a pretty straightforward example of a Realist approach—and one used to deny a right to AI systems because they simply are not enough like human beings.[18]

Interestingly, the Court does indicate the possibility in the future of there being sufficiently complex machines that are creative in a way required for them to genuine authors, but by contrast, present-day AI systems are simply not creative in way that is required.

> Of course, the Creativity Machine does not represent the limits of human technical ingenuity when it comes to artificial intelligence. Humans at some point might produce creative non-humans capable of responding to economic incentives. Science fiction is replete with examples of creative machines that far exceed the capacities of current generative artificial intelligence. . . . This court's job, by contrast, "is to apply the statute as it is written," not to wade into technologically uncharted copyright waters and try to decide what "might 'accord with good policy.'"

There will be time enough for Congress and the Copyright Office to tackle those issues when they arise.[19]

Perhaps it is noteworthy that there is not a discussion, much less a definition, given of "creativity." We are simply left with the idea that creativity is human endeavor, though perhaps the issue will be revisited when the need presents itself.

Arguably, though, the issue *has* already presented itself, for if the Court's view is not just based on an unapologetic bias in favor of humans, and what,

16 *Thaler v. Perlmutter*, No. 23-5233 (D.C. Cir. 2025) <https://media.cadc.uscourts.gov/opinions/docs/2025/03/23-5233.pdf>

17 Machines, it is claimed, do not own property. They also lack traditional human lifespans, family members, domiciles, nationalities, *mentes reae,* and signatures. These are all cited as evidence that machines cannot be authors. *Thaler v. Perlmutter* 11fn.

18 "The point, instead, is that the current Copyright Act's text, taken as a whole, is best read as making humanity a necessary condition for authorship under the Copyright Act." *Thaler v. Perlmutter*, 13.

19 *Thaler v. Perlmutter*, 20.

in fact, is driving the copyright dismissal is a belief that AI systems are not creative, or not creative enough, that seems to warrant an account of what creativity is in the first place. Put otherwise, how will we know when we're faced with AI systems that are truly creative if we don't have even a rough articulation of what that standard looks like? If the strategy is to (at least someday, possibly) assimilate AI's art output to that of humans, with an eye to deciding whether the former sufficiently resembles the latter (in the way we would expect a Realist perspective on AI personhood to seek), then presumably, we need a stable characterization or a paradigmatic instance of human creativity so that we can assess the distance from it and AI productions, as they continue to evolve and become more sophisticated as the technology does.[20]

Legalisms, Interests, and AI

To wrap up this Appendix, I wish to address a cluster of remaining issues and puzzle that arise for AI personhood when we connect our earlier discussion of interests and personhood. The argument there was that legal persons have rights in virtue of having interests that can be harmed. So, if AI is to be a person, it must have rights, and for it to have rights, it must have interests. As we've seen throughout our discussions, for a Weak Legalist, the only feature required of would-be persons is the having of interests, though, crucially, those interests do not have to resemble human interests in any way. More generally, they need not be the interests of the kind had by sentient beings. We saw this in our discussion of corporations, in the broad sense of the term, that have non-sentient-based interests. If AI, however, should turn out to *not* have interests, that would seem to disqualify it as potentially a legal person for even Weak Legalism. With reasons to be skeptical of present-day weak-AI's claims to be sentient, much less to be rational entities, we would presumably find many Realists also opposed to the personhood of AI. Perhaps all that would be left would be what I've called Strong Legalism, which, as we saw in the Introduction, seemingly allows for anything to be a legal personhood.

I've claimed earlier in this Appendix that *instrumental* justifications for AI personhood look to line up with Legalist views, with both contrasting with a view that focuses on the *intrinsic* properties of the AI and purported

20 We might even imagine the production of an aesthetic variant of the famous Turing test for determining whether a computer program is intelligent. Such a project is spelled out in Costanza Cenerini, Flavio Keller, Giorgio Pennazza, Marco Santonico and Luca Vollero, 'Artistic Turing Test: The Challenge of Differentiating Human and AI-Generated Art' (2025) 6 *Computers in Human Behavior: Artificial Humans* 100209 <https://doi.org/10.1016/j.chbah.2025.100209> For background on the Turing test and its relevance to contemporary AI, see M Mitchell, 'The Turing Test and Our Shifting Conceptions of Intelligence' (2024) 385(6710) *Science* <https://doi.org/10.1126/science.adq9356>

similarities with humans. Connecting Legalism and an instrumental justification is, perhaps, overly simplistic, however, and now is the appropriate place to be more precise. Recall that Strong Legalism, as I've introduced the term, claims that personhood is not tied to any feature thought essential to humans. It goes even further in holding that there are no constraints at all on what could be legal person, hence, its claim that anything could potentially be a legal person. In other words, Strong Legalism is indifferent to *any* intrinsic features of a purported legal person, arguing that what matters is not what the entity is like but rather whether making that entity a legal person is the best (or at least a better) way to accomplish a social-economic-political goal than the alternatives.

Weak Legalism, on the hand, *does* care about what an entity is like, though, like Strong Legalism, it doesn't matter whether there is a resemblance with human beings. Instead, we might say, there is an intrinsic feature that is relevant for the Weak Legalist, and that is whether a proposed entity *has interests*. The ability to be harmed (or made better) is the intrinsic feature that matters, as it were, if something is to be a legal person. For the Weak Legalist, the possibility of AI being a person depends on whether it has that intrinsic feature—whether it has interests that can be harmed.[21] Figure 4.1 demonstrates how these competing theories relate to each other, with emphasis on the degree of difficulty in being recognized as a rights-bearer.

All things considered, I am inclined to say that AI systems lack interests. If that is right, we find ourselves with a bit of a puzzling conclusion if such an AI system were embedded in a functioning robot.[22] As a physical object, the robot—the mechanical parts—might very well have interests, especially given the very low bar I've put in place for holding things have interests. It is in the interest of the robot not to fall into a vat of acid. Doing so would harm the robot, make it worse off—hence, its interest in avoiding such a vat. Apparently, for the Weak Legalist, the hardware has interests, could have rights, and thus could be a legal person, while the AI system, lacking interests, would not be a person—even though, ironically, it is the AI system itself in which we were expecting (hoping?) to find a legal person.[23]

21 We saw in Chapter 3 that the notion of intrinsic value is notoriously complicated and the subject of ongoing debates. So, too, for "intrinsic properties." A helpful primer can be found in 'Intrinsic vs. Extrinsic Properties' <https://plato.stanford.edu/entries/intrinsic-extrinsic/>

22 Or, perhaps, in a self-driving car.

23 We can generate the puzzle even without the robot. The mere fact that an AI system is implemented in a physical medium, be it silicon chips or neurons, would seemingly have the same consequences—of the physical system having interests and, thus, potential personhood, while the AI system without interests falls short.

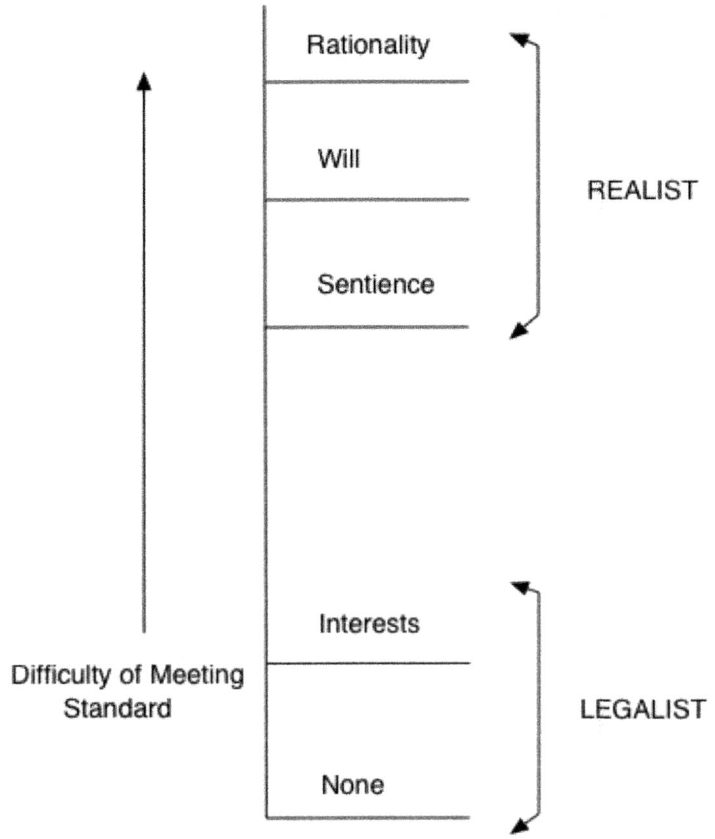

Figure 4.1 Threshold for Rights

But why would I claim, or suspect, that AI systems lack interests? Put simply, thinking clearly about what an AI system actually is gives us reason to think it is just not the kind of thing that can be made better or worse. Admittedly, it is not easy to fully capture the essence of an AI system with a simple definition, and there are multiple ways we could try. Let us lead up to that by considering related items that also don't seem to be capable of being made better or worse. For instance, a mathematical formula. It can be altered. It can be changed from being true to being false. But that would not, I claim, be a change that makes it better or worse. It would just be different.

Newton's second law is that F=ma, where the net force on a body is equivalent to its mass times its rate of acceleration. As an equation, we, of course, are free to change it—and can imagine a (fictional) world where his predecessors had different permutations as they struggled to get it right. Let us instead then say that

$$F=ma^2$$

Have we changed the equation? Or have we simply offered a different one? Perhaps the answer doesn't matter, but what does is that we now have an equation that does not accurately describe the relationship between mass, acceleration, and force. And most importantly, I claim, our new equation hasn't made the initial equation itself better or worse. It might be worse for predicting the behavior of physical objects. It may be worse aesthetically for us. But it is hard to see how the equation itself is worse, even if we grant that it has undergone a change.

Consider now an algorithm. Described very simply, an algorithm is a set of instructions for completing a task. A simple example might be to do what's called a "linear search," which would involve searching for a particular item among a manifold. The instructions, or recipe, if you like, would tell us to search the first item in our manifold to see if the desired object is present. If it is, the task is completed. If it is not, we move to the second item. If the object is present there, the search is completed. If not, we continue, etc.

As with the formula example, we could alter that algorithm, modifying the instructions, modifying the order, modifying how it determines what is the next item to be searched, etc. But again, I think it is hard to find room to say the algorithm itself has been made better or worse. It might be better at achieving a task we want it to—thus, better for us or, alternatively, worse for us if we've set it out to achieve a task. It might do worse, for instance, or take more time, in completing a task. Thus, its utility might be changed. But I still don't see room for saying it itself is better or worse *full stop*. The modification is just a change. Not a change for better or worse.

The point of all this, as a lead-up, is to reckon with the idea that AI systems themselves are just very complicated algorithms.[24] If we understand AI as a very complicated algorithm or series of algorithms, those, again, don't seem to be the kinds of things that can made better or worse for the AI *in itself*. Here, some care is required because there *is* a sense in which it is better for an AI to be trained on a large data set as opposed to a small one or to have its data normalized or not to have functionless code added to it. And the coding can improve or worsen an AI and its output. Do those not count

24 I am radically simplifying here. Unlike traditional algorithms, AI systems, such as LLMs, do not rely on a preset determined 'recipe' or set of instructions but instead make (probabilistic) predictions based on the connections established through their training using huge sets of data to create a model. For more on the exact workings of an LLM, see I Blank, 'What Are Large Language Models Supposed to Model?' (2023) 27 *Trends in Cognitive Sciences* 987.

as examples of making an AI better or worse, thus demonstrating that it has interests in the thin sense I've developed earlier?

Here, I'm inclined to say those statements are elliptical, and a full statement would make clear *who* those changes are better/worse for, or for *what* they are better or worse. An LLM trained on a small data set is worse than one trained on a larger set, all things being equal. But it is better for the *users* of the AI system, not the system itself. And it is better for the developer of the AI system who wishes to increase sales of their AI system. An AI system trained in poorly assembled data sets doesn't fulfill its programmed function as well for the *user* nor as well as it could for the *developer* of software that builds on the original AI using poor data.

If we understand an AI as a complex series of algorithms, any changes for those algorithms may be said to produce a new algorithm. Were we to try and get a bit more precise about AI systems, we would talk about how the weightings between inputs and outputs are changed. The point remains the same, I think. For it isn't clear in what way the changes are good or worse for the algorithm itself. Likewise, if we think of the AI as an abstract mathematical formula that describes the complex relationship between inputs and outputs, making changes to it might be better for the user, *qua* user experience, or better for the developer, *qua* market share. But again, it is far from clear that it is a change for better or for worse for the AI code/algorithm/equation. Here, it seems that AI systems are not the kinds of things that can undergo changes for the better or for the worse. At most, they can simply change.

If all of this is correct, or in the neighborhood, then AI systems don't have interests. For they are not the kinds of things that can be made better or worse. Without interests, they no longer qualify for personhood even under a Weak Legalist understanding. Of course, there is room to disagree in my descriptions of what exactly an AI system is, and perhaps my characterizing them as simply complicated algorithms or complicated mathematical formulas is misguided as well as overly simplistic. That might give us a way to redescribe them as capable of being made better or worse and, hence, as having interests. But I'll leave the discussion here.

Finally, though, I mention an apparent puzzle I alluded to before, one that arises if we agree that AI systems, as mathematic abstractions or algorithms, lack interests. There is the low bar I've set for having interests. That has entailed the possibility for physical objects to have interests. So in the name of finding a way to give natural objects interests, we might have to agree that the physical implementation of an AI system, in robots, in silicon, rubber, metal, etc., can themselves be harmed. But they can be harmed only if they have interests. And, thus, we have a puzzle upon our hands, akin to a new mind-body problem, as it were. The "mind" would be AI software system that doesn't have rights or interest, while the robot, as "body," does. Any duties we might have would seem directed at the body, the robot, as that is the element that has interests, though the AI system itself does not.[25] If nothing else, this would be rather ironic, given the initial thought we may have had that if anything is to be deserving of rights, and put duties onto others, it would be the AI system. After all, *that* is the part that we begin thinking is the element most like us.

As for the implications for RoN, we may find ourselves having to rethink exactly how the connection between AI systems and their physical implementations works when we turn back to ecosystems. Is an ecosystem more like a software program that "runs" on the material elements found in a given place—the biological and geological elements that make up the hardware? If so, would the appropriate interests track primarily with the physical, as in the AI case, or with the intangible "system" of functions that "run" on the hardware of the materials of the ecosystem? If so, does that mean that any obligation we have are to protect the physical structures and materials of an ecosystem, not the processes and "emergent" functions of the ecosystem that "flow" from the hardware, the physical? These are but some of the many, many questions that arise as we confront the challenge of expanding our notions of personhood, rights, and interests beyond the bounds of the human paradigm, the very problematic we are embarked on as we seek to go beyond the anthropocentrism of environmental law and policy, and of law most generally.

References

Baeyaert J, 'Beyond Personhood: The Evolution of Legal Personhood and Its Implications for AI Recognition' (2025) 2025 *Technology and Regulation* 355, 374 <https://doi.org/10.71265/ssvg8a97>

Blank I, 'What Are Large Language Models Supposed to Model?' (2023) 27 *Trends in Cognitive Sciences* 987.

Bory P, Natale S and Katzenbach C, 'Strong and Weak AI Narratives: An Analytical Framework' (2025) 40 *AI & Society* 2107 <https://doi.org/10.1007/s00146-024-02087-8>

Cenerini C, Keller F, Giorgio P, Santonico M and Vollero L, 'Artistic Turing Test: The Challenge of Differentiating Human and AI-Generated Art' (2025) 6 *Computers in Human Behavior: Artificial Humans* 100209 <https://doi.org/10.1016/j.chbah.2025.100209 >

Chopra S and White F, *A Legal Theory for Autonomous Artificial Agents* (University of Michigan Press 2011) 158.

Coeckelbergh C, 'Robot Rights? Towards a Social-Relational Justification of Moral Consideration' (2010) 12 *Ethics and Information Technology* 209.

Gellers JC, *Rights for Robots: Artificial Intelligence, Animal, and Environmental Law* (Routledge 2021).

25 A rich discussion of the many issues and challenges that arise when we think about robots, their nature, and possible legal standing can be found in JC Gellers, *Rights for Robots: Artificial Intelligence, Animal, and Environmental Law* (Routledge Press 2021).

Guadamuz A, 'Do Androids Dream of Electric Copyright? Comparative Analysis of Originality in Artificial Intelligence Generated Works' in J-A Lee, RM Hilty and K-C Liu (eds), *Artificial Intelligence and Copyright Law* (Oxford Academic Press 2021) <https://doi.org/10.1093/oso/9780198870944.003.0008>

'Intrinsic vs. Extrinsic Properties' <https://plato.stanford.edu/entries/intrinsic-extrinsic/>

Jumper J, Evans R, Pritzel A, et al., 'Highly Accurate Protein Structure Prediction with AlphaFold' (2021) 596 *Nature* 583 <https://doi.org/10.1038/s41586-021-03819-2>

Mitchell M, 'The Turing Test and Our Shifting Conceptions of Intelligence' (2024) 385(6710) *Science* <https://doi.org/10.1126/science.adq9356>

Novelli C, Floridi L, Sartor G and Teubner G, 'AI as Legal Persons: Past, Patterns, and Prospects' (2024) 6 *Journal of Law and Society* <https://doi.org/10.71265/ssvg8a97>

Solum L, 'Legal Personhood for Artificial Intelligences' (1992) 70 *North Carolina Law Review* 1231 <https://scholarship.law.unc.edu/nclr/vol70/iss4/4>

Thaler v. Perlmutter, No. 23-5233 (D.C. Cir. 2025) <https://media.cadc.uscourts.gov/opinions/docs/2025/03/23-5233.pdf>

Yanisky-Ravid S and Liu XJ, 'When Artificial Intelligence Systems Produce Inventions: The 3A Era and an Alternative Model for Patent Law' (2018) 39(6) *Cardozo Law Review* 2215.

Chapter 5

Rights of Nature, Laponia, and New Zealand

In this chapter and the next, I seek to explore how a Legalist account of personhood and rights might play out in the Arctic, by way of a hypothetical example. I choose to develop this hypothetical case for several reasons. When I introduced the notion of legal visibility, I noted that, in general, this is a way the law has of signaling the special status of a certain class of entities, and with that status comes a distinct set of norms for treatment of those entities. Most of this book has focused specifically on two kinds of visibility—the giving of rights to an entity and the granting of legal personhood to that entity. But I also noted that there is a degree of visibility that runs shy of rights and personhood. I characterized this level as one that signals a particular object is worthy of protections, though those protections do not amount to the kinds given by rights. An example would be the designation of an area of land for which environmental impact assessments are necessary before that land can be used. Alternatively, we have examples where there are strict prohibitions of the use of land for economic or military purposes. These would be protections regarding the land (or ecosystem, river, etc.) but without the designation of rights or personhood for the land (ecosystem, etc.) itself. RoN aims to create stronger protections for various environmental entities, by giving them rights and/or personhood. In my terminology, RoN seeks to increase the level of legal visibility for such entities.

In order to see the implications of a change to the visibility of an environmental entity, I proceed with an entity that has the lowest level of visibility and investigate the differences were its status to change as we move "up the ladder" of visibility. What, that is, might be different were that natural entity to go from not having rights to having rights of its own? Or were it to go from being a legal *object* to being a legal *subject*, a person of law? How would that change the kinds of protections the entity might have? How might a change in legal status change our understanding and appreciation of that entity? And, following the general line of inquiry of this book, if we are to imagine giving that natural entity its own legal personality, what advantages would there be in understanding its personhood along the lines of Legalism, as opposed to Realism?

DOI: 10.4324/9781003740186-6

Those questions provide one motivation for my decision to focus on Laponia, a legally protected area in northern Sweden. Laponia has been used for thousands of years by the Sami Peoples for their livelihood, culture, and religion, which adds another layer of significance. As we saw earlier in discussion of Tănăsescu, the project of giving nature its own rights and personhood has been developed by some as part of the larger goal to further the rights of self-determination for Indigenous Peoples, including their rights to the lands they have used since time immemorial.[1] This is in addition to drawing on Indigenous Peoples' understanding of the natural world as potential antidote to anthropocentric Western views, which often view nature merely as an exploitable resource.

Finally, the selection of Laponia offers another opportunity—namely, to explore in a tentative way what RoN might look like and the difference it might make in a part of the world that largely has not pursued a RoN approach to the environment on anywhere near the scale it has in other parts of the world. As noted on the website ecojurisprudence.org, which tracks RoN and related initiatives worldwide, there are relatively few RoN initiatives in the Arctic. Even among those few, there is an extremely limited range of attempts to give nature legal personhood.[2] By way of explanation for the relatively few examples in the Arctic, particularly in contrast to, say, Latin America and Central Europe, we might make use of Mihnea Tănăsescu's observation that "the rights of nature are not about nature, but rather about the political relations between different groups of people (some of whom might understand themselves as people of particular lands)."[3] If that is correct, then the relative lack of initiatives in the Arctic is certainly not attributed to a lack of concern about the environment but rather reflects that the political relations in the Arctic differ from other places in the world. It may be that the politics of, say, Indigenous Peoples' rights to self-determination play out in a way in the Arctic that make RoN less favorable to Indigenous Peoples there, or simply that environmental concerns are not framed in the same legal and political manner as elsewhere. A full answer that goes beyond mere speculation is more apt for analysis by, say, political scientists than it is international lawyers pursuing legal theories about nature, rights, and personhood.

1 E O'Donnell, A Poelina, A Pelizzon and C Clark, 'Stop Burying the Lede: The Essential Role of Indigenous Law(s) in Creating Rights of Nature' (2020) 9(3) *Transnational Environmental Law* 403 <https://doi.org/10.1017/S2047102520000242>

2 Eco Jurisprudence Monitor <https://ecojurisprudence.org/dashboard/?map-style=political>. Though there are various ways of defining the Arctic, I intend the eight countries at or above the Arctic circle: Iceland, United States of America, Canada, Norway, Sweden, Finland, Denmark (via Greenland), and Russia.

3 M Tănăsescu, 'The Rights of Nature as Politics' in D Corriga and M Oksanen (eds), *Rights of Nature: A Re-Examination* (Routledge 2021) 69.

The downside, or rather one among various possible downsides of this proposed exploration, is that it is speculative and not tied down by actual initiatives to change the legal status of Laponia. Still, I think there is value in that openness. Many of the actual RoN initiatives have been explored in depth, and there are established and entrenched lines of argument and interpretation. The Laponia case, because it is hypothetical, perhaps gives us a chance to see some opportunities and possibilities that disappear or are closed off once the political, social, and legal battles begin. My hope is that there is a degree of freedom in exploring the possibilities for change before different sides become entrenched in their views and oppositions.

Laponia, a region in northern Sweden covering approximately 9,400 square kilometers, was designated a World Heritage Site in 1996. The area satisfies multiple criteria from different categories for inclusion as a heritage site, being considered a "mixed" site worthy of protection for its combination of both natural and cultural heritage and values, including the rights of the Sámi People, who have inhabited the area for thousands of years. In what follows, I move from how the site is presently governed, which involves a host of legal protections for both those natural and cultural places, to how it might look should it be granted legal rights of its own. In the next chapter, I advance the discussion by considering not just legal rights but also legal personhood and how that might change the legal landscape in Laponia. Finally, I examine how a Legalist interpretation of that personhood proves preferable to a Realist one.

What makes this an interesting project, among other things, is the similarity between the Laponia site and the cases of New Zealand discussed previously, in particular, the legal personhood that has been accorded the Whanganui River (Te Awa Tupua) and the Te Urewera forest.[4] For while Laponia has not been granted legal personhood, in both regions, the desire to provide environmental protections are intimately connected with protecting the rights of the Indigenous Peoples who live in the areas: Māori in New Zealand and Sámi in Sweden. So, too, as we'll see, there are important similarities in the governance structures in the New Zealand regions and Laponia. Indeed, I will draw upon the similarities to examine how governance and protection of both nature and the Sámi People's rights might look were Laponia granted legal personhood.

To be clear, what I'll be offering is *not* a contrast between law as it is (*lex lata)* compared with how it ought to be (*lex ferenda)*. I am not in a position to speak of what is best for the Sámi People of Sweden and am not qualified to speak on their behalf. Instead, this will be an exercise in seeing what

4 In late January 2025, legal personhood was also granted to Taranaki Maunga (Mt Taranaki) in New Zealand, 'New Zealand Mountain Gets Same Legal Rights as a Person' <https://www.bbc.com/news/articles/czep8gg5lx4o>

is possible in the law, or to coin a phrase, *lex possibilis*. For it is worth asking what might be different when it comes to the protection of nature and Sámi rights and culture were Laponia to be granted legal personhood. Finally, as mentioned, I will use this an exercise to examine how a Legalist understanding of personhood might fare in a particular case, again contrasting it with the Realist understanding developed in earlier chapters.

Background to Laponia and Swedish Law

Prior to its declared status as a World Heritage Site, 95% of Laponia was already a protected area through its four national parks (Muddus, Sarek, Padjelanta, and Stora Sjöfallet) and two nature reserves (Sjaunja and Stubba). Importantly, Laponia also includes significant areas that are ongoing sites for Sámi reindeer herding. After a lengthy process of negotiation, the Laponia site began to be managed by Laponiatjuottjudus ("Laponia Administration") in 2013, which is an administrative board comprised of representatives from the Sámi community, Sweden's Environmental Protection Agency (SEPA), the Norrbotten County Administrative Board, and representatives of the local municipalities of Jokkmokk and Gällivare.[5] Sámi members comprise a majority of the administrative board, and decisions are made via consensus.[6]

According to UNESCO, Laponia meets multiple criteria which allows for its designation as a "mixed site."[7] In particular, Laponia is said to qualify in virtue of its meeting of five of UNESCO's criteria, in particular: iii, v, vii, viii, and ix. The criteria are divided into two categories, *cultural* and *natural*, and are as follows, where I quote at length how Laponia has been said to meet each criterion, drawing on UNESCO's formulation and language.[8]

A. Cultural Criteria

Criterion (iii): To bear a unique or at least exceptional testimony to a cultural tradition or to a civilization which is living, or which has disappeared.

"The Laponian Area bears exceptional testimony to the tradition of reindeer herding, and is one of the last and unquestionably the largest and best preserved examples of an area of transhumance, a practice once widespread

5 'Laponia—The Management' <https://laponia.nu/en/the-management/>

6 O Stjernström, A Pashkevich and D Avango, 'Contrasting Views on Co-management of Indigenous Natural and Cultural Heritage—Case of Laponia World Heritage Site, Sweden' (2020) 56(e4) *Polar Record* 1 <https://doi.org/10.1017/S0032247420000121>

7 'UNESCO—The Criteria for Selection' <https://whc.unesco.org/en/criteria/>

8 'Laponian Area' <https://whc.unesco.org/en/list/774/>. Used with permission under License, Creative Commons Attribution-ShareAlike 3.0 IGO <https://creativecommons.org/licenses/by-sa/3.0/igo/deed.en>

in northern Europe and which dates back to an early stage in human economic and social development."

Criterion (v): To be an outstanding example of a traditional human settlement, land-use, or sea-use which is representative of a culture (or cultures), or human interaction with the environment especially when it has become vulnerable under the impact of irreversible change.

"The Laponian Area is an outstanding example of traditional land-use, a cultural landscape reflecting the ancestral way of life of the Sámi People people based around the seasonal herding of reindeer."

B. Natural Criteria

Criterion (vii): To contain superlative natural phenomena or areas of exceptional natural beauty and aesthetic importance.

"The property exhibits a great variety of natural phenomena of outstanding beauty. The snow-covered mountains in Sarek and Sulidälbmá are not only magnificent to see but are a textbook of glacial-related geomorphology. The large alpine lakes in Padjelanta, with the mountain backdrop on the Swedish/Norwegian border are of exceptional beauty. The extensive Rapa Valley provides a total contrast with the alpine areas. Particularly noteworthy is its very active delta area, surrounding cliffs and rocky outliers with sheer faces plunging into the delta. The existence of the Sámi culture ranging from the traditional birch and turf kata to contemporary cabins adds to the aesthetic value of the property."

Criterion (viii): To be outstanding examples representing major stages of earth's history, including the record of life, significant on-going geological processes in the development of landforms, or significant geomorphic or physiographic features.

"The nominated area contains all the processes associated with glacial activity such as U-shaped valleys, moraines, talus slopes, drumlins, presence of large erratics and rapidly flowing glacial streams. It has excellent examples of ice and frost action in a tundra setting including formation of polygons and an area of spectacularly collapsing and growing palsa mounds. Glacial rivers originating in the snowfields continue to cut through bedrock. Large unvegetated areas illustrate the phenomenon of weathering. The property also contains a record of humans being part of these ecosystems for 7,000 years.

Criterion (ix): To be outstanding examples representing significant on-going ecological and biological processes in the evolution and development of terrestrial, fresh water, coastal and marine ecosystems and communities of plants and animals.

"The vast mire complex of Sjávnja/Sjaunja is the largest in Europe outside Russia. This area is virtually impenetrable by human beings except during winter. The Laponian area has primeval coniferous forest with dating indicating ages as old as 700 years. Natural succession continues here unimpaired."

So while Laponia is unique in the kind of place it is, we can also see that its governance is appropriately unique as well. It derives from a distinctive arrangement of Sámi influence and interplay with the more traditional Swedish governmental structures. Most broadly, Sweden is comprised of 21 counties and 290 municipalities, with each county having a County Administrative Board tasked with implementing and administrating federal policies.[9]

The County Administrative Board of Norrbotten County, which is represented on Laponiatjuottjudus, is responsible for oversight of environmental policy goals for that region, as well as the management of nature reserves, national parks, and mountains within Laponia. Authority for its role in Laponia derives from various legislative acts, including the Environmental Code, the Cultural Heritage Act, the Reindeer Husbandry Act, the Hunting Act, and the Off-Road Driving Act. It is those pieces of legislation that gives the County Administrative Board authority to implement Sweden's Environmental Code.[10]

In virtue of Laponia being a mixed site, it requires protection of both its physical environment as well as its cultural heritage. It is, thus, crucially important that the Swedish National Heritage Board (known as RAÄ, or Riksantikvarieämbetet) is involved in oversight, as it is the central authority for the latter, particularly in its leading role in cultural heritage initiatives, including conservation and archaeological commission activities, as well as being responsible for distributing of government grants.[11]

In addition to supporting the county administrative boards, RAÄ is responsible for the periodic reporting to UNESCO of the state of the World Heritage Site. Further cooperation between RAÄ and Laponiatjuottjudus is envisaged on issues that specifically concern Laponia.[12]

Another relevant actor here is the Sámi Parliament, which is both a government agency and an elected Sámi parliament with an overall mission to monitor issues relating to Sámi culture in Sweden. The Sámi Parliament is also the central administrative authority and the government's expert authority with responsibility for issues relating to the reindeer industry.[13]

9 'Swedish Environmental Law: An Introduction to the Swedish Legal System for Environmental Protection' (2017) *Report* 6790 <https://www.divaportal.org/smash/get/diva2:1421472/FULLTEXT01.pdf>

10 'Laponiaprocessen: A New Management Organization for the World Heritage Site of Laponia,' 14. <https://www.slu.se/centrumbildningar-och-projekt/centrum-for-naturvagledning/naturvagledning/verktyg/planering/informationsvagledning-laponia/> Machine Translation by Google.

11 ibid, 16–17.

12 ibid, 17.

13 'Sami Parliament' <https://www.sametinget.se/om-sametinget>

As a World Heritage Site that includes both natural and cultural heritage, and given its location within natural parks and nature reserves, there is a constellation of legal regimes relevant to the management of the site. These include the Reindeer Husbandry Act (1971); the Swedish Environmental Code (1999), particularly Chapter 7 ("Protection of Areas," which covers both natural parks, natural reserves, and cultural reserves"); Chapter 8 ("Special Provisions Concerning the Protection of Animal and Plant Species"); the Historic Environment Act (1988); the Planning and Building Act (2010); and the Forestry Act (1979).

The Reindeer Husbandry Act (the "Act") guarantees the exclusive and collective right of reindeer husbandry to the Sámi People, although only a member of a Sámi reindeer herding village (sameby) is able to exercise that right.[14] Nine of these villages are located in Laponia, including both mountain and forest reindeer herding areas.

> The Act provides a host of explicit and various implicit rights for Sámi, including "basic rights to graze the reindeer and move between different pasture areas, rights to build structures and buildings necessary for the management of the reindeer, including a family house, and rights to take timbers for such buildings and structures, as well as rights to take fuel and material for handicraft. The right to hunting and fishing is also important. There might be additional sub-rights not explicitly mentioned in the Act, such as rights to gravel and other pits, as well as rights to gather lichen for feeding the reindeer when enclosed."[15]

Even though Laponia is a protected site, there is still a range of permitted activities for locals and tourists, such as nature-based tourism, including hiking, cross-country skiing, observing wildlife, and participating in guided tours. Such activities are allowed but regulated in a way that seeks to prevent the disruption of ecosystems and local cultures, including, especially, the reindeer herding rights.[16] As expected, there are also various prohibited activities, such as unauthorized camping, damaging vegetation, disturbing wildlife, driving off designated roads, littering, collecting plants or rocks without permission, large-scale organized events without approval, and any activity that significantly alters the landscape, with special consideration given to the traditional reindeer herding practices.[17]

14 'Reindeer Husbandry in Sweden—Rights to Own Reindeer and "Sameby"' <https://icr.arcticportal.org/sweden?lang=en&start=1>

15 Christina Allard, 'Two Sides of the Coin- Rights and Duties: The Interface Between Environmental Law and Saami Law Based on a Comparison with Aoteoaroa/New Zealand and Canada' PhD Thesis (Luleå University of Technology 2006) 357.

16 'Laponia Adventures' <https://laponiaadventures.com/>

17 'How to Behave within the Laponia World Heritage Site' <https://laponia.nu/en/how-to-behave-within-the-laponia-world-heritage-site/>

As for Sweden's Environmental Code, which provides broad effect over all of Sweden, a number of goals and values are present, which seek an often elusive balance between sustainable development on the one hand and environmental protection on the other, as well as preservation of the environment for recreational purposes.

> The purpose of this Code is to promote sustainable development which will assure a healthy and sound environment for present and future generations. Such development will be based on recognition of the fact that nature is worthy of protection and that our right to modify and exploit nature carries with it a responsibility for wise management of nature.[18]

An example of the various values enshrined in the Code can be seen in its description of "natural reserves."

> A land or water area may be declared a nature reserve by a county administrative board or a municipality for the purpose of preserving biological diversity, protecting and preserving valuable natural environments or satisfying the need of areas for outdoor recreation.
>
> Any area that is needed for the purpose of protecting, restoring or establishing valuable natural environments or habitats for species that are worthy of preservation may also be designated a nature reserve.[19]

Here, we find a mix of values, as one would expect, given Laponia's World Heritage status, as well as how Sweden understands its connection to nature most generally. On the one hand, there is language that does speak to the direct value of nature itself. The language of "intrinsic value," "ultimate worth," and the like is missing, to be sure. Still, there seems to be at least a gesture towards recognizing that Laponia and nature in general has a value beyond the mere instrumental, which we find specifically mentioned in describing nature as "worthy of protection." Coupled with this is the straightforward anthropocentric valuing of nature that we've discussed before, where nature is valued as a means towards meeting human needs. Phrases such as "our right to modify and exploit nature" are clear indicators of this view. Finally, the mentioned "responsibility for wise management" seems to be aimed at ensuring that nature is not exploited to a degree that robs future generations of its utility, though there does not seem to be explicit language aimed at intergenerational equity.

18 'The Swedish Environmental Code' 9 <https://www.government.se/contentassets/be5e4d4ebdb4499f8d6365720ae68724/the-swedish-environmental-code-ds-200061/>

19 ibid, 37.

Finally, the particular governance structure of Laponia is worth mentioning, as it represents an important break from other protected areas in Sweden in the makeup of its members, in its consensus model for decision-making, and also in the parallels I later examine that it bears to governance structures in regions of New Zealand.

As mentioned, administration via Laponiatjuottjudus is managed by a board representing a host of interested parties in Sweden, with particular emphasis given to the voice of Sámi. For many, this represents a break from Swedens's traditional policy of leaving all decisions in the hands of the County Administrative Boards. In contrast, this board reflects a willingness to allow and encourage local voices to be heard and to share in the administration of conservation efforts.[20]

A primary challenge here for Laponia is that it is not just environmental protection that is sought but also protection of cultural heritage, both its history and its ongoing usage. The management board is intended to give priority to local concerns and needs. The policies articulated in national legislation, such as the Environmental Code, are extremely broad and need to be given particular effect through administrative offices and decisions. In this case, it is the Laponiatjuottjudus that is tasked with implementing those policies.[21]

The Function and Potential of Rights

As seen, Laponia is presently governed by a host of legal regimes, ones that seek to preserve both the natural and cultural values found there. What those protections don't do is give Laponia rights in and for itself. As a general matter, we can ask how the language of rights might change matters were it employed on a behalf of something that, while having legal protections, doesn't have the status of being a rights-bearing entity. What, we might ask, does the assignment of rights offer? RoN, after all, is motivated by the notion that giving rights *to* nature itself as opposed to merely having rights *regarding* nature is an important legal distinction. In my terminology, what changes when we move from the lowest level of visibility to the next higher one?

As discussed in earlier chapters, interest and will theories help us specify a domain of beings that have rights based on features deemed necessary and/or sufficient for being a rights-holder. Rights are intended to reflect the special character qualified beings have, with rights working to protect those specified features, whether that is the ability to make rational choices or, in the case of the interest theory, to protect the basic welfare of something.

20 Laponiaprocessen: A New Management Organization, 3.
21 ibid, 11.

On a different understanding of rights, something we have not yet addressed, rights provide legal reasons, which can then justify various actions. More precisely, rights provide preemptory reasons for doing (or refraining) from some action. Their preemptory status means that such rights will count as more important than other reasons one might have for doing (or not doing) a given action. In the hands of this view's most famous defender, Ronald Dworkin, having a right to something is said to outweigh other considerations, typically utilitarian ones that may factor into reasoning about whether a given action is permissible.

> Individual rights are political trumps held by individuals. Individuals have rights when, for some reason, a collective goal is not a sufficient justification for denying them what they wish, as individuals, to have or to do, or not a sufficient justification for imposing some loss or injury upon them. That characterization of a right is, of course, formal in the sense that it does not indicate what rights people have or guarantee, indeed, that they have any. But it does not suppose that rights have some special metaphysical character, and the theory . . . therefore departs from older theories of rights that do rely on that supposition.[22]
>
> So a claim of political right is a claim to a trump over the general welfare for the account of a particular individual. When someone claims a right of free speech, for example, he claims that it would be wrong for the state to prohibit him from speaking on some matter even if the general welfare would be improved by preventing him from speaking. The hypothetical nature of the claim is important. We emphasize the special injustice of torture, for example, when we speak of a right against torture, because we claim that torture would be wrong even if it were in the general interest.[23]

So, for example, having a right to free speech is supposed to, in Dworkin's language, "trump" utilitarian considerations that focus on the benefits of censuring speech, say, the prevention of various harms to others. Having a right to free speech means that there is consideration in favor of free speech that count as qualitatively more important than benefits that comes from censuring it. That is what rights do. Thus, were a particular river to have a right to flourish, that becomes a reason for leaving the river unspoiled, a reason that outweighs the benefits that might come from damming the river for hydropower. Because rights-talk carries such preemptory weight, it signals where our priorities lay.

Stone's famous essay on whether trees should be granted legal standing can help us develop this point. In a section entitled "Do We Really Have to

22 Ronald Dworkin, *Taking Rights Seriously* (Harvard University Press 1977) 7.
23 ibid, 431.

Put it That Way?" he asks whether the language of rights he draws upon is really necessary, or whether we could more simply get by, say, expanding a list of protections and enforcement mechanisms. Could we not, he wonders, simply extend the range of people who are competent to intervene on nature's behalf to curtail pollution, or simply strengthen various environmental protections?[24] By way of reply, Stone offers a host of reasons, many of which are directly relevant to our inquiries. I quote him here at length, numbering the different points I take him to be making about the importance of affording nature its own rights and discuss them in turn.

> Introducing the notion of something having a "right" (simply speaking that way), brings into the legal system a flexibility and open-endedness that no series of specifically stated legal rules like R1, R2, R3 . . . Rn can capture. Part of the reason is that [1] "right" (and other so-called "legal terms" like "infant," "corporation," "reasonable time") have meaning—vague but forceful—in the ordinary language, and the force of these meanings, inevitably infused with our thought, becomes part of the context against which the "legal language" of our contemporary "legal rules" is interpreted. . . . In the case of such vague rules, it is context—senses of direction, of value and purpose—that determines how the rules will be understood, every bit as much as their supposed "plain meaning." In a system which spoke of the environment "having legal rights," judges would, I suspect, be inclined to interpret rules such as those of burden of proof far more liberally from the point of the environment. [2] There is, too, the fact that the vocabulary and expressions that are available to us influence and even steer our thought. Consider the effect that has had by introducing into the law terms like "motive, "intent," and "due process." These terms work a subtle shift into the rhetoric of explanation available to judges; with them, new ways of thinking and new insights come to be explored and developed. In such fashion, judges who could unabashedly refer to the "legal rights of the environment" would be encouraged to develop a viable body of law—in part simply through the availability and force of the expression. [3] Besides, such a manner of speaking by courts would contribute to popular notions, and a society that spoke of the "legal rights of the environment" would be inclined to legislate more environment-protecting rules by formal enactment . . .
>
> If my sense of these influences is correct, then a society in which it is stated, however vaguely, that "rivers have legal rights" would evolve a different legal system than one which did not employ that expression, even if the two of them had, at the start, the very same "legal rules" in other respects.[25]

24 C Stone, *Should Trees Have Standing?: Law, Morality, and the Environment* (Oxford University Press 2010), 22.

25 ibid, 22–23.

Points [1] and [2] are about the power of the language of rights to shape the law. For instance, we can imagine an influence on the legal rules relevant to the burden of proof a mining company might have to provide to show that its intended activities are not going to harm a rights-bearing natural region it intends to develop. So, too, Stone points out that having a foundation in rights for natural entities provides encouragement for new laws to be developed and legal reasoning about the status of right-holding nature to be sharpened and refined. His final point, [3], speaks to the potential for influencing popular discourse and to even shape eventual legislation.

In [3], Stone can perhaps be seen as gesturing towards what, following Cass Sunstein and others, we may call the "expressive power" of law.[26] For law not only structures our behavior through encouragement and deterrence. It also functions to embed our values and shape our norms. For instance, in many States, suicide is illegal. The deterrence factor here is likely minimal, as someone intent on suicide is unlikely to worry about the legality or possible prosecution of their actions. But the stated illegality serves another purpose, namely, to articulate society's views about the value and importance of human life. Or to use an example from Sunstein:

"[M]any people who endorse capital punishment would not be much moved by evidence that capital punishment does not deter people from committing crimes. Their primary concern is the symbolic or expressive content of the law, not aggregate murder rates."[27]

Not only does law play a role in articulating and promulgating a society's values, it is also "designed to affect social norms and in that way ultimately to affect both judgments and behavior."[28] Taking Stone and Sunstein together, we may say that the power of granting rights to nature, assuming as per earlier arguments that it *is* the kind of thing that can have rights in the first place, is the power such rights have to shape the norms and values we accord nature. Granting rights to nature would change the legal landscape in so far as new legal rules may be put in place, including the kinds of changes Stone mentions in [1] and [2] earlier. It holds the potential to shape our norms and behaviors, including, following Dworkin, giving overriding reasons to refrain from certain activities that threaten nature and for taking pro-active measures on nature's behalf. RoN, thus, has the potential to effect widespread change not only in hallowed court settings but also in a wider and more diffuse manner. As David Boyd has optimistically put it, RoN can be a legal revolution that saves the world.[29]

26 Cass Sunstein, 'On the Expressive Function of Law' (1996) 144 *University of Pennsylvania Law Review* 2021 <https://papers.ssrn.com/sol3/papers.cfm?abstract_id=2622561>

27 ibid, 2023.

28 ibid, 2025.

29 David R Boyd, *The Rights of Nature: A Legal Revolution That Could Save the World* (ECW Press 2017). A dissenting view can be found in Julien Bétaille, 'Rights of Nature: Why It

I believe the potential *expressive power* from laws or judicial decrees giving rights (and personhood) to nature is an underappreciated source for change, one that has received little attention in the literature. A rare exception to this neglect can be found in Baard, where he writes of how:

> Law has the possibility of shaping experiences and a productive relation between law and ethics can benefit justifications of RoN. By "rights" law has the potential to cohere with intrinsic values of both anthropocentric and non-anthropocentric kinds, the unique resources in nature, and the urgency of recurrent threats to the environment. "Rights" shape experiences and relations and has unique potential to express urgency of mitigating threats to environmental entities carrying substantial value.[30]

Giving rights to nature, to use terminology I introduced early on, would elevate its legal visibility. To be sure, laws that protect the environment, say, by requiring environmental impact assessments before development can occur on a given piece of land, do signal to the public that the environment matters. Just the same, giving rights to nature signals, in a more powerful way, how nature matters—and thus, I claim, elevating its legal visibility.

Consider this thought experiment. Imagine a society where nature has rights and personhood of its own and that these legal standings have been in place for decades. Now imagine a new regime comes to power and strips its Constitution and various statutes of any mention of nature's rights. Nature is reduced to being an object, and though there will still be conservation laws in place, no longer will nature have rights or personhood. Now consider the fallout of these changes. Leave aside material consequences of these changes, such as the increase in pollution that may follow. Have there been other changes? In particular, have there been changes in the message, meanings, and intentions of laws that mention ecosystems, the environment, rare species, etc.? I submit that the intuitive answer is yes and that nature has undergone a kind of harm irrespective of how much changes in rates of pollution or sea-level rise or temperature gradients. Nature has lost something, it seems, in being demoted from that second and third degree of visibility.[31]

Might Not Save the Entire World' (2019) 16(1) *Journal for European Environmental & Planning Law* <https://doi.org/10.1163/18760104-01601004>

30 P Baard, 'Rights of Nature Through a Legal Expressivist Lens: Legal Recognition of Non-Anthropocentric Values' (2025) 28 *Ethical Theory Moral Practice* 551, 556 <https://doi.org/10.1007/s10677-024-10479-4> (Licensed under CC BY 4.0).

31 Readers will note that I am in the neighborhood of claiming that nature has here undergone an *expressive harm*. I'm less interested in asserting and defending that particular way of putting it than I am in simply mobilizing the intuition that there is an expressive meaning and force that comes with talk of rights and personhood. A classic paper on expressive harm is,

If that is correct, then consider the reverse—where a society goes from not having nature with rights and personhood of its own to having those stipulations and understandings in law. By parity of reasoning, I think we'd again say something significant has happened in this case as well. That, I maintain, helps us see the expressive power contained in the language of rights (and personhood).

Speaking of nature's rights expresses a kind of importance that nature has and may very well lead the public to think of other occasions where rights were deployed as a way of effecting social change, or reorienting people's priorities, and giving a message about the importance of the beings who have rights. A particular example of this would be the Civil Rights Movement in the United States in the 1950s and 60s, where segregation was outlawed and where the rights of African Americans were reaffirmed and strengthened, including as well, via the Voting Rights Act of 1965, the elimination of voter suppression tactics.[32] The Civil Rights Act of 1964 outlawed segregation in public accommodations, schools, and the workplace based on race, religion, or sex.[33]

With that as a model for how law's expressive content (i.e. the purpose and intent contained as message signaling the normative content of law) can shape society, we can transpose it to understand the potential expressive power that would come from granting rights to nature. The endorsement of nature as a rights-bearing entity would elevate its status and thereby help to shape and modify how we understand and value nature.

Rights for Laponia

In the abstract, we have seen multiple ways in which granting rights to nature can go beyond what is accomplished by simply making certain regions of nature legally protected areas. In the case of Laponia, it is, no doubt, a good thing that it is presently viewed as a protected site, one which permits certain activities and prohibits others. But the *grounds* for those protections matter. My hypothesis is that putting those prohibited areas instead under the auspices of *rights* is a stronger foundation for prohibiting those actions. It is a way of saying that the balance of interests is different. Without grounding those prohibitions in actual rights that Laponia itself has, it is easy to see that the balance of conflicting interests can very well end up giving way to actions that are harmful to Laponia.

Elizabeth S Anderson and Richard H Pildes, 'Expressive Theories of Law: A General Restatement' (2000) 148(5) *University of Pennsylvania Law Review* 1503.

32 Signed into law by President Johnson, the Act is codified in the US Code, primarily at 52 U.S.C. §§ 10101 et seq.

33 The Civil Rights Act of 1964: Pub. L. 88–352, 78 Stat. 241, enacted July 2, 1964. It is codified at 42 U.S.C. § 2000a et seq.

For instance, the interests of tourists are taken into consideration when determining which areas in Laponia are accessible and how they can be used. Sweden grants "rights of access" to nature, though those rights are limited in Laponia.[34] But even such limited access seems inherently in tension with the interests Laponia itself may be said to have in being free from interference. Granting legal rights to Laponia itself could be seen as giving reasons to leave it alone, reasons that perhaps override interests the general public may have in visiting and exploring Laponia. That's the upshot of something like Dworkin's view about the preemptory status of rights.

Yet to this imagined view about Laponia's rights, we face an immediate complexity. After all, it is not only tourists that want to make use of Laponia. It is also the Sámi People who use Laponia for their reindeer grazing lands. And Laponia has already been granted protection for the cultural heritage of Sámi. Wouldn't giving Laponia its own rights, say, to flourish and thrive, immediately jeopardize the rights of the Sámi? If Laponia is to have rights against intrusion, wouldn't that exclude the Sámi as much as it does tourists looking for a picturesque viewing spot?

As noted in the quotation from Feinberg in Chapter 3, in a diverse society, there will always be the challenge of balancing competing interests of different parties. How might we address complicated questions about the balancing of interests—of Laponia, of tourists, and of the reindeer herding Sámi? Here, I think it is worth considering the rights of the Sámi People as an Indigenous People and what their rights look like under international law. Grounded rights of self-determination here might be said to offer more weighty reasons than tourists' interests in experiencing relatively untouched wilderness. And perhaps that can provide reasons for allowing Sámi uses of Laponia that override the interests of mere tourists.

Sweden's endorsement of the Universal Declaration on the Rights of Indigenous Peoples (UNDRIP)[35] can perhaps underwrite this difference in the weight accorded to different parties seeking to make use of Laponia. Although UNDRIP is only a declaration, reflecting at least some aspects of customary international law, such as the rights of peoples' self-determination,

34 'How to Behave Within Laponia.'

35 Article 3 of UNDRIP speaks specifically of the right of self-determination. "Indigenous peoples have the right to self-determination. By virtue of that right they freely determine their political status and freely pursue their economic, social and cultural development." UN General Assembly, *United Nations Declaration on the Rights of Indigenous Peoples: Resolution/ Adopted by the General Assembly*, A/RES/61/295, 2 October 2007 <https://www.refworld.org/legal/resolution/unga/2007/en/49353>

It is worth noting, however, that Sweden is not a party to the legally binding ILO 169. International Labour Organization (ILO), C169—Indigenous and Tribal Peoples Convention, 1989, No.169, 27 June 1989, <https://www.refworld.org/legal/agreements/ilo/1989/en/19728>

it does, in its own way, signal that in Sweden's view, the rights of the Sámi People *do* count.[36] Granting that the Sámi People do have various rights under the banner of self-determination, we might say, following the logic of Dworkin again, that their rights, as rights, put their interests in a position to trump the interests of others. Much depends, of course, on which rights are granted under UNDRIP and what Sweden's overall position is on the legal force of UNDRIP.

On the question of access and use of Laponia, relevant to our present discussion, various articles of UNDRIP are germane, though I'll focus only on Articles 26 and 32, which state:

"Article 26:

1. Indigenous peoples have the right to the lands, territories and resources which they have traditionally owned, occupied or otherwise used or acquired.
2. Indigenous peoples have the right to own, use, develop and control the lands, territories and resources that they possess by reason of traditional ownership or other traditional occupation or use, as well as those which they have otherwise acquired.
3. States shall give legal recognition and protection to these lands, territories and resources. Such recognition shall be conducted with due respect to the customs, traditions and land tenure systems of the indigenous peoples concerned."

"Article 32:

1. Indigenous peoples have the right to determine and develop priorities and strategies for the development or use of their lands or territories and other resources."
2. States shall consult and cooperate in good faith with the indigenous peoples concerned through their own representative institutions in order to obtain their free and informed consent prior to the approval of any project affecting their lands or territories and other resources, particularly in connection with the development, utilization or exploitation of mineral, water or other resources."

While I've mentioned the competing interests of Sámi, Laponia, and tourists, it is worth exploring another clash of interests where Laponia having rights of its own might make a difference. Here, it might be useful to examine these articles in light of the controversial Kallak mine project, particularly

36 See S James Anaya, 'The UN Declaration on the Rights of Indigenous Peoples: Towards Re-Empowerment' <https://www.jurist.org/commentary/2007/10/un-declaration-on-rights-of-indigenous-2/?>

as the issue of granting mining concessions bears on Laponia's status as a UNESCO site.

A substantial deposit of iron ore had been discovered in 2013, in a region approximately 34 km southwest of Laponia, and Sweden declared the area of national interest for its deposits of valuable substances. In 2020, Sweden submitted to UNESCO information on a proposed mining project, Kallak, that is in the vicinity of Laponia. It was then suggested by UNESCO that rigorous impact assessments be done to estimate the potential impact on Laponia and its OUV (outstanding universal value), prior to any approval of the mining project. After initial impact assessments were completed, UNESCO expressed its concerns that as of 2022, there still had not been sufficient attention in the EIA (environmental impact assessment) to potential damage to Laponia, in particular, the impacts on "water, energy and the land use requirements related to the project."[37]

Moreover, there were concerns that the mining concession process failed to adequately consult the Sámi People, as required by UNDRIP.

> Furthermore, the significant concerns raised by the SRs regarding a lack of consultation with relevant indigenous peoples including the Sámi village which would be directly impacted by the proposed development, are of deep concern. Noting also the objections expressed by indigenous peoples and local communities in the State Party's decision, it is important to reiterate that any decision taken in relation to cultural practice relation to World Heritage values should ensure appropriate consultation and clear evidence that free, prior and informed consent has been obtained, in line with the United National Declaration on the Rights of Indigenous Peoples (UNDRIP) and the Operational Guidelines.[38]

On this latter issue, of the requirements of providing free, prior, and informed consent, UNDRIP is clear. In addition to Article 32, earlier, Article 19 states:

> States shall consult and cooperate in good faith with the indigenous peoples concerned through their own representative institutions in order to obtain their *free, prior and informed consent* before adopting and implementing legislative or administrative measures that may affect them.[39]

37 'Laponian Area.'

38 United Nations Special Rapporteurs Report: AL SWE 1/2022 <https://spcommreports.ohchr.org/TMResultsBase/DownLoadPublicCommunicationFile?gId=27057>

39 UNDRIP, Article 19.

Sweden's position on such matters of consultation is worth mentioning, as it seems to view consent as *not* required but rather only that *consultation*, *notification*, *publicity*, and the holding of meetings be done, as per the Swedish Environmental Code.

> They shall also *consult* private individuals who are likely to be affected and must do so in good time and to an appropriate extent before submitting an application for a permit and preparing the environmental impact statement that is required.[40]
>
> When an environmental impact statement is prepared in a case or matter relating to environmentally hazardous activities or water operations, *notification* thereof shall be published together with the notice of the application referred to in chapter 19, section 4 and chapter 22, section 3 . . . If an environmental impact statement has been prepared for another case or matter concerning an activity or measure that is likely to have a significant environmental impact, notification thereof shall be published. If an application has been submitted, notification of the environmental impact assessment shall be published together with notification of the application. The application and the environmental impact statement shall thereafter be made available to the public, which shall be given the opportunity to comment on them before the case or matter is dealt with.[41]

Sweden's view on UNDRIP itself reveals a fairly weak reading of what Sweden believes UNDRIP requires with respect to free, prior, and informed consent.

> According to the State party, the meaning of the concept of free, prior and informed consent, as expressed, for instance, in article 19 of the United Nations Declaration on the Rights of Indigenous Peoples, has been disputed. That Declaration is not legally binding and the concept does not entail a collective right of veto . . . [and] consent may not be required when a limitation on indigenous peoples' rights is considered to be necessary and proportional in relation to a valid State objective.[42]

Now, of course, it is hard to know how enshrining legal rights of Laponia itself might specifically alter the legal terrain, especially given Sweden's inclination to downplay the legal significance of UNDRIP. On the other hand, however, *were* Laponia to have legal status itself, beyond just legal

40 Swedish Environmental Code, 32.

41 ibid, 34.

42 *Ågren et al. v. Sweden* (CERD/C/92/D/54/2013) <https://wolverine-opossum-zznd.squarespace.com/s/Ronnbacken-SC-Open-Letter-062021.pdf>

protections under the various relevant statutes and codes, that might very well reshape the relationship of Sámi and the lands in Laponia. *If* Laponia had legal rights of its own, then perhaps there would have been a different and stronger foundation for legal action against the Kallak mine for violation of Laponia's rights. That's a far stronger legal position to take, given as we've seen the importance and value of rights. *If* Laponia had rights of its own, then violations of those rights would perhaps be of a different order than simply infringing against sections in the Environmental Codes, especially if there had been specific rights granted to Laponia against exploitation, for instance.

There are perhaps some nascent reasons to be optimistic about the possibility that Sweden is open to rethinking its stance on rights, particularly on Sámi land rights. For instance, in the Nordmaling case, the Swedish Supreme went beyond deciding land rights based on standards of immemorial prescription, instead agreeing that Sámi customary law provided justification for Sámi claims to use winter pasture grounds on private property.[43] The mere recognition of the legal standing of Sámi customary law, used to ground claims that Sámi, in fact, had rights to those lands, despite being privately owned, perhaps signals a shift towards recognizing the legitimacy of not only Sámi claims per se but perhaps more deeply on the foundations that Sámi customary law has. True, this is still shy of granting outright ownership to the reindeer herders, much less to granting Laponia legal rights on its own. Still, given that legal change is slow, movement in the direction of recognizing Sámi claims should not be minimized.

Notice that the preceding discussion has spoken of Laponia as having its own set of rights. That is different, as I've observed through earlier chapters, from speaking of Laponia as having legal personality, though, of course, the two are intimately related. As discussed, both would be ways of increasing Laponia's *legally visibility* from its present status. Giving Laponia specific rights might offer a different kind of legal protection than that provided through Sweden's Environmental Code, for instance. But, as I've suggested, the legal personhood of Laponia might be an even stronger position. For often what is implied in granting the legal personhood to a heretofore non-person is that the newly legally visible person can be said to have a much wider range of rights than can be found in cases where only a few specific rights have been granted.

43 Nordmaling Case No. T 4028–07, decided on April, 27, 2011. Specific references to the case can be found in Christina Allard, 'The Swedish Nordmaling Case' (2011) 2 *Arctic Review on Law and Politics* 225; Christina Allard, 'Some Characteristic Features of Scandinavian Laws' in Christina Allard and Susann Funderud Skogvang (eds), *Indigenous Rights in Scandinavia: Autonomous Sámi Law* (Routledge Press 2015) 59; Bertil Bengtsson, 'Reforming Swedish Sámi Legislation' in Christina Allard and Susann Funderud Skogvang (eds), *Indigenous Rights in Scandinavia: Autonomous Sámi Law* (Routledge Press 2015) 67.

For instance, were Laponia a legal person, it might be able to own property. Should it be able to be possible for it to *own itself*—to have the kind of legal protection a person has in their own body—that would seemingly strengthen the land rights that Laponia has. This will be explored further in the next chapter. For now, it is worth noting that depending on how those rights are articulated, and how Sámi are connected with those particular rights of self-ownership, we would very likely have a very different legal landscape when it comes to the relationship Sámi have to their native lands. Ownership, instead of mere use-rights, would perhaps be on the table in a way that has not been. Ownership of Laponia, whether by Laponia itself, by local Sámi herders, or a creative combination of the two, very well might strengthen the legal protections Laponia has against potential mining projects and unwanted tourist intrusions, not to mention further enshrining the environmental protections in more legally powerful local legal regimes.

All in all, what the preceding discussion shows in part is that there would be consequential legal changes were Laponia to be granted rights of its own, ones I've only begun to sketch. On other side, however, the possibility of those changes faces challenges from both the domestic and international legal side. Sweden's entrenched policies regarding the rights of the Sámi People and the substantial weight borne by Sweden's environmental policies among other legal regimes makes it hard, as a political matter, to imagine radical reform. On the international front, Sweden's seeming downplaying of the significance of UNDRIP poses its own challenges. As one writer has put it, "However, it remains to be seen if a Swedish government will take so much trouble for such an end. At least for the time being, the chances of a Sámi law reform seem doubtful."[44]

Laponia and New Zealand

As a final point in the chapter, I wish to begin examining what legal personhood for Laponia might look like by some drawing parallels with New Zealand's famous examples of the granting of legal personhood to Whanganui River (Te Awa Tupua) and the Te Urewera forest. I will continue examining implications of Laponia's personhood in the next chapter, but a few closing points are worth making here.

As mentioned in Chapter 4, the personhood examples in New Zealand are examples of localized natural regions, specifically designated as legal persons.[45] In both cases, the specified areas are granted "all the rights, powers,

44 Bengtsson, 'Reforming Swedish Sámi,' 65.

45 The Te Awa Tupua Act specifies the Te Awa Tupua to be a legal person. "Te Awa Tupua is a legal person and has all the rights, powers, duties, and liabilities of a legal person. The rights, powers, and duties of Te Awa Tupua must be exercised or performed, and responsibility for

and duties of a legal person." Crucially, in each case, guardianship to exercise those rights and duties is granted to an oversight board.[46]

Important for our purposes is that, like Te Awa Tupua and Te Urewera, Laponia is a particular designated area, which, in the previous chapter, I described as a preferable way of describing "nature," compared with a generic or universal understanding of "Nature." Furthermore, the governance in the New Zealand examples is done by an appointed set of guardians, charged with acting on the interests of the specified regions. This parallels the guardianship model seen in Laponia, where it is to be managed by an administrative body, Laponiatjuottjudus.

As highlighted by Kauffman, a particularly important role for such administrative boards, whether we are talking about the New Zealand examples, or extending our discussion to include the management of Laponia, is the forward-looking aspect of that governance. Unlike some of the examples described earlier in Ecuador, via the Aquepi River case, the focus is not on finding legal remedies for harms that have already been committed against a particular natural entity. Instead, it is pro-active, seeking to anticipate potential harms and working to develop planning for the well-being and thriving of that natural entity. That is to say, the administrative bodies are charged with acting in the best interests of the natural body that has been granted legal personhood. In the case of Laponia, we might imagine how Laponiatjuottjudus could have its administrative authority extended not only as a representative body for managing a legally protected area, as it is now, but

its liabilities must be taken by Te Pou Tupua on behalf of, and in the name of, Te Awa Tupua, in the manner provided for in this Part and in Ruruku Whakatupua—TeMana o Te Awa Tupua." Te Awa Tupua (Whanganui River Claims Settlement) Act, 2017 Article 14 <https://www.legislation.govt.nz/act/public/2017/0007/latest/whole.html>

The Te Urewera Act 2014, Article 11, grants legal personhood to the Te Urewera forest. "Te Urewera is a legal entity, and has all the rights, powers, duties, and liabilities of a legal person. However, (a) the rights, powers, and duties of Te Urewera must be exercised and performed on behalf of, and in the name of, Te Urewera—(i) by Te Urewera Board; and (ii) in the manner provided for in this Act; and (b) the liabilities are the responsibility of Te Urewera Board." <https://www.legislation.govt.nz/act/public/2014/0051/latest/dlm6183705.html>

46 As Kauffman explains, "In order to effectuate the Act, an Office of human guardians was established with all the powers necessary to protect and preserve the River. . . . Not only are these Guardians charged with protecting the intrinsic value of the River, but they are able to assert landowner rights for those lands vested in the River. . . . These Guardians, Advisors, and Strategists act as an integrated support system that is embedded in a collaborative governance body tasked with managing the protected ecosystem, and that 'management' is characterized as protecting the River's 'health and well-being' as a natural being. In New Zealand's model, decision-making takes the ecosystem's rights and interests into consideration and requires collaborative planning among the stakeholders." Kauffman C, 'Guardianship Arrangements in Rights of Nature Legal Provisions' in Anthony Zelle (ed), *Earth Law: Emerging Ecocentric Law—A Guide for Practitioners* (Wolters Kluwer 2020) <http://ebookcentral.proquest.com/lib/uoregon/detail.action?docID=6460079>

also with further legal powers, namely, as a governing body charged with acting in the legal interests of Laponia itself, with my imagined twist that Laponia is itself a legal person. Notice as well that for Te Awa Tupua, there is the legal status allowing the governing body to "assert landowner rights for those lands," which might be a useful model for addressing the thorny question of land rights I mentioned earlier for the Sámi People.

A final parallel worth mentioning is that for Laponia, even in its present state of being merely a legal object, the interests of the Sámi, who rely on it for its reindeer grazing pastures, are intwined with management of Laponia as a place of natural heritage. That, after all, is why Laponia was created as a *mixed* heritage site, one that encompasses both natural and cultural heritage. This looks to parallel how the mentioned areas in New Zealand also incorporate the needs and interests of both the natural landscapes as well as the interests of the local Māori populations. As the Act helpfully puts it:

"Te Awa Tupua is a spiritual and physical entity that supports and sustains both the life and natural resources within the Whanganui River and the health and well-being of the *iwi*, *hapū*, and other communities of the River."[47]

References

'2020 Conservation Outlook Assessment' <https://worldheritageoutlook.iucn.org/explore-sites/laponian-area>

Ågren et al. v. Sweden (CERD/C/92/D/54/2013) <https://wolverine-opossum-zznd.squarespace.com/s/Ronnbacken-SC-Open-Letter-062021.pdf>

Allard C, 'Two Sides of the Coin-Rights and Duties: The Interface Between Environmental Law and Saami Law Based on a Comparison with Aoteoaroa/New Zealand and Canada' PhD Thesis (Luleå University of Technology 2006).

Allard C, 'The Swedish Nordmaling Case' (2011) 2 *Arctic Review on Law and Politics* 225.

Allard C, 'Some Characteristic Features of Scandinavian Laws' in C Allard and S Funderud Skogvang (eds), *Indigenous Rights in Scandinavia: Autonomous Sámi Law* (Routledge Press 2015) 59.

Anaya SJ, 'The UN Declaration on the Rights of Indigenous Peoples: Towards Re-empowerment' <https://www.jurist.org/commentary/2007/10/un-declaration-on-rights-of-indigenous-2/?>

Anderson ES and Pildes RH, 'Expressive Theories of Law: A General Restatement' (2000) 148(5) *University of Pennsylvania Law Review* 1503.

Baard P, 'Rights of Nature Through a Legal Expressivist Lens: Legal Recognition of Non-Anthropocentric Values' (2025) 28 *Ethical Theory Moral Practice* 551 <https://doi.org/10.1007/s10677-024-10479-4> (Licensed under CC BY 4.0).

Bengtsson B, 'Reforming Swedish Sámi Legislation' in C Allard and S Funderud Skogvang (eds), *Indigenous Rights in Scandinavia: Autonomous Sámi Law* (Routledge Press 2015) 67.

47 Te Awa Tupua Whanganui River Settlement 2014, Article 13 <https://www.whanganui.govt.nz/About-Whanganui/Our-District/Te-Awa-Tupua-Whanganui-River-Settlement>

Bétaille J, 'Rights of Nature: Why It Might Not Save the Entire World' (2019) 16(1) *Journal for European Environmental & Planning Law* <https://doi.org/10.1163/18760104-01601004>

Boyd DR, *The Rights of Nature: A Legal Revolution That Could Save the World* (ECW Press 2017).

The Civil Rights Act of 1964: Pub. L. 88–352, 78 Stat. 241, enacted July 2, 1964. It Is Codified at 42 U.S.C. § 2000a et seq.

Dworkin R, *Taking Rights Seriously* (Harvard University Press 1977) 7.

Eco Jurisprudence Monitor <https://ecojurisprudence.org/>

'How to Behave Within the Laponia World Heritage Site' <https://laponia.nu/en/how-to-behave-within-the-laponia-world-heritage-site/>

International Labour Organization (ILO), C169—Indigenous and Tribal Peoples Convention, 1989, No.169, 27 June 1989 <https://www.refworld.org/legal/agreements/ilo/1989/en/19728>

Kauffman C, 'Guardianship Arrangements in Rights of Nature Legal Provisions' in A Zelle (ed), *Earth Law: Emerging Ecocentric Law—A Guide for Practitioners* (Wolters Kluwer 2020) <http://ebookcentral.proquest.com/lib/uoregon/detail.action?docID=6460079>

'Laponia Adventures' <https://laponiaadventures.com/>

'Laponia—The Management' <https://laponia.nu/en/the-management/>

'Laponian Area' <https://whc.unesco.org/en/list/774/> (Creative Commons Attribution-ShareAlike 3.0 IGO CC BY-SA 3.0 IGO).

'Laponiaprocessen: A New Management Organization for the World Heritage Site of Laponia' 14 <https://www.slu.se/centrumbildningar-och-projekt/centrum-for-naturvagledning/naturvagledning/verktyg/planering/informationsvagledning-laponia/> Machine Translation by Google.

'New Zealand Mountain Gets Same Legal Rights as a Person' <https://www.bbc.com/news/articles/czep8gg5lx4o>

Nordmaling Case No. T 4028–07, decided on 27 April 2011.

O'Donnell E, Poelina A, Pelizzon A and Clark C, 'Stop Burying the Lede: The Essential Role of Indigenous Law(s) in Creating Rights of Nature' (2020) 9(3) *Transnational Environmental Law* 403 <https://doi.org/10.1017/S2047102520000242>

'Reindeer Husbandry in Sweden—Rights to Own Reindeer and "Sameby"' <https://icr.arcticportal.org/sweden?lang=en&start=1>

'Sami Parliament' <https://www.sametinget.se/om-sametinget>

Stjernström O, Pashkevich A and Avango D, 'Contrasting Views on Co-management of Indigenous Natural and Cultural Heritage—Case of Laponia World Heritage Site, Sweden' (2020) 56(e4) *Polar Record* 1 <https://doi.org/10.1017/S0032247420000121>

Stone CD, *Should Trees Have Standing?: Law, Morality, and the Environment* (Oxford University Press 2010).

Sunstein C, 'On the Expressive Function of Law' (1996) 144 *University of Pennsylvania Law Review* 2021 <https://papers.ssrn.com/sol3/papers.cfm?abstract_id=2622561>

The Swedish Environmental Code' 9 <https://www.government.se/contentassets/be5e4d4ebdb4499f8d6365720ae68724/the-swedish-environmental-code-ds-200061/>

'Swedish Environmental Law: An Introduction to the Swedish Legal System for Environmental Protection' (2017) *Report* 6790 <https://www.diva-portal.org/smash/get/diva2:1421472/FULLTEXT01.pdf>

Tănăsescu M, 'The Rights of Nature as Politics' in D Corriga and M Oksanen (eds), *Rights of Nature: A Re-Examination* (Routledge Press 2021) 69.

Te Awa Tupua (Whanganui River Claims Settlement) Act, 2017 <https://www.legislation.govt.nz/act/public/2017/0007/latest/whole.html>

'UNESCO—The Criteria for Selection' <https://whc.unesco.org/en/criteria/> (Creative Commons Attribution-ShareAlike 3.0 IGO CC BY-SA 3.0 IGO).

UN General Assembly, United Nations Declaration on the Rights of Indigenous Peoples: Resolution/Adopted by the General Assembly, A/RES/61/295, 2 October 2007 <https://www.refworld.org/legal/resolution/unga/2007/en/49353>

United Nations Special Rapporteurs Report: AL SWE 1/2022 <https://spcommreports.ohchr.org/TMResultsBase/DownLoadPublicCommunicationFile?gId=27057>

Chapter 6

The Rights and Personhood of Laponia

The previous chapter began an ascent, as it were, from viewing Laponia as a legally protected object to an imagined status where it had rights of its own. As sketched in that chapter, my hope was that imagining Laponia with its own rights would offer a stronger legal foundation for the protection of the natural and cultural values that motivated UNESCO to declare Laponia a World Heritage Site in the first place.

Having examined the power of something having rights of its own—in both a general way and more specifically in the case of Laponia—I also began to sketch the next step in our ascent and how Laponia as a legal person might further change the legal landscape. In this chapter, I aim to develop this further, looking at how legal personhood, as opposed to what I consider merely the recognition of the legal rights of something like Laponia, provides the potential for an even deeper modification of its legal regime. To be sure, even imagining Laponia having rights of its own, perhaps akin to what we've seen in cases from Colombia and Ecuador, would represent a huge legal transformation. But as the bulk of this book is geared towards understandings of the possibilities for *personhood*, I wish to push the imagined scenario even further. What then would be different were Laponia given its own personhood? Finally, to round out the discussion began in Chapter 1, what would a Legalist understanding of Laponia's personhood look like? Would it offer possibilities that were otherwise not available if we remained within a Realist framework?

The view I will explore and defend in this chapter holds that, ultimately, a Legalist rendering of Laponia's imagined personhood would allow for an easier and more respectful acknowledgment of the worldview of the Sámi People, who have lived, worked, and survived there since time immemorial. I will suggest, put somewhat differently, that inherent in a Legalist reading of personhood is a greater potential for tolerance of divergent metaphysics and values and that accommodating Laponia along Legalist lines opens the possibility for a greater fit between the traditional Western understandings of what a human person is (in the non-legal sense) and a Sámi conception of persons (again, in the non-legal sense). That is, Legalism's inherent

DOI: 10.4324/9781003740186-7

disposition towards tolerance allows for an easier juxtaposition of two radically different worldviews, making it easier in theory to accommodate Sámi conceptions within a Western-legal context.[1] Furthermore, I will argue that a Legalist understanding of Laponia's personhood offers greater possibilities for the establishment and eventual development and evolution of the *rights* that could accrue to Laponia.

In order to reach these conclusions, I will explicate what I understand to be a Sámi conception of human beings, nature, culture, and related notions, seeking to understand from the inside, as it were, how the Sámi People themselves understand matters and pointing out along the way that the very distinctions so familiar to the Western eye—distinctions between nature and culture, for instance, must themselves be reconceptualized to truly appreciate the differences in these two distinct worldviews.[2]

The Power of Personhood

I've argued in previous chapters that there is a hierarchy of legal visibilities. As it stands in Laponia, there are various legal protections in place, ones that have regard for its well-being and for the interests of the Sámi population. But Laponia itself, at present, is still merely a legal object. Imagining it being given rights of its own, akin to cases elsewhere in the world, would give it a different kind of standing—as a subject of rights. I want to suggest now, as gestured at several times, that we should distinguish two kinds of relationships between having rights and being a legal person, what I spoke of as indirect and direct bestowals of personhood.

On this "indirect" relationship, we have examples from Ecuador and Colombia, where Nature (or a particular portion of nature, say, a river) is granted particular rights, perhaps the right to exist, to flourish, or to be restored. Given the tight conceptual connection between being a rights-bearer and being a legal person—where in Chapter 1, I argued that to be a legal

1 It is far from uncontroversial that it is a good thing to construct a scenario where Sámi worldviews are put side by side with Western-legal concepts. I will not try to argue for the merits of this juxtaposition. Nor for the possibility of a genuine legal pluralism. Instead, recognizing that we already have a pluralist society in Sweden, one that combines both Sámi and Western values, I will suggest that a Legalist understanding of Laponia's personhood makes for an easier accommodation of those two very different worldviews than any Realist construal could.

2 I intentionally use the indefinite article and speak of "a" Sámi understanding of persons and the like, for I do not assume that there is simply one, monolithic Sámi view, especially given that there are many different populations spread across the Arctic, from Norway to Russia, that self-identify as Sámi. Just as it would be a mistake to speak of the "Asian perspective" on the Arctic, so, too, I think it a mistake to assume that all Sámi worldviews will be identical. The scholarship I draw on, unless indicated otherwise, focuses on the Sámi People of northern Sweden.

person just is to be the bearer of rights and duties—being granted particular rights is to imply that the rights-bearer in question is a legal person. After all, non-legal persons don't have rights of their own. So, again, were we to grant particular rights to something, say, a right for Laponia to exist, then I think we have, in an important sense, granted the legal personhood of Laponia. But, and this is the key, that personhood has been granted in an *indirect* or *implied* manner. This, I believe, is in important contrast with cases such as we've seen in New Zealand, where a particular region or river or ecosystem is *directly* granted personhood by being *explicitly* made a legal person. I think this is a distinction with an important difference.

When we find cases of Nature (or "nature") being directly or explicitly granted legal personhood, we typically find language that speaks of it as having all the various rights and duties of a *natural* person. My view is that this *explicit* recognition of personhood (in contrast with the *implied* version where personhood flows from particular rights being bestowed) is the zenith of legal visibility and represents a form of legal recognition of an importantly different order. One reason for this higher status lies in the different scope of rights that come from these two different routes to personhood. Being granted particular rights to exist or flourish is obviously limited compared with being granted "all the rights of a natural person." In addition, and this is the deeper point, that wider range of rights allows for the development and evolution of rights in a way that signals the potential for even greater legal and legislative change.

Recall the discussion of Stone from Chapter 4, where I listed his three ways of how the language of nature's rights could be particularly important. In what I labeled as reason [2], I quoted Stone as saying that "In such fashion, judges who could unabashedly refer to the 'legal rights of the environment' would be encouraged to develop a viable body of law—in part simply through the availability and force of the expression."[3] Stone is surely correct that in giving nature rights (and with it personhood, either indirectly or directly), a new body of law may eventually be created. What I think is also worth examining is not just the potential for new law but also for an ongoing, evolving, and developing understanding of what those laws actually mean. That is to say there is the potential for development and evolution in our *interpretation* of those laws. This potential, I claim, is even greater when personhood has been explicitly granted, as in New Zealand–type scenarios.

An example from international law can serve to explain the manner in which our interpretation of law and rights itself can develop. From there, we can illustrate that how personhood is generated adds another layer of

3 CD Stone, *Should Trees Have Standing? Law, Morality, and the Environment* (3rd edn, Oxford University Press 2010) 22.

potential for such developments, with direct or explicit granting carrying a greater possibility of evolution than mere implicit granting. First things first.

Consider Article 27 of 1966 International Covenant on Civil and Political Rights (ICCPR):

> In those States in which ethnic, religious or linguistic minorities exist, persons belonging to such minorities shall not be denied the right, in community with the other members of their group, to enjoy their own culture, to profess and practice their own religion, or to use their own language.[4]

From *our* perspective decades later, the then perhaps groundbreaking status of Article 27 may seem rather simplistic. Article 27 simply seems to be directing States to not engage in discrimination. Put differently, the only rights members of minority groups seem to have are so-called "negative rights." No burden is placed on States to effect any action to further those rights, and the rights given are simply rights to not be interfered with—what Berlin famously spoke of as "freedom from," as opposed to "freedom to."[5] But an understanding of what anti-discrimination requires has developed since then. Consider the Human Rights Committee, the UN treaty body in charge of monitoring states' compliance with the ICCPR, in its General Comment No. 23, from *1994*. It notes that anti-discrimination measures require *positive* actions from States. States have a duty to do more than simply *not* engaging in discriminatory behavior.

> Although article 27 is expressed in negative terms, that article, nevertheless, does recognize the existence of a "right" and requires that it shall not be denied. Consequently, a State party is under an obligation to ensure that the existence and the exercise of this right are protected against their denial or violation. Positive measures of protection are, therefore, required not only against the acts of the State party itself, whether through its legislative, judicial or administrative authorities, but also against the acts of other persons within the State party.[6]

We can find further developments in understandings of what anti-discrimination requires by noting, from *2009*, General Comment 32 on the Committee on the Elimination of All Forms of Racial Discrimination (CERD),

4 International Covenant on Civil and Political Rights (adopted 19 December 1966, entered into force 23 March 1976) 999 UNTS 171 art 31(1).

5 Isaiah Berlin, 'Two Concepts of Liberty' in Henry Hardy (ed), *Liberty* (2nd edn, Oxford University Press 2002).

6 UN Human Rights Committee (HRC), CCPR General Comment No. 23: Article 27 (Rights of Minorities), CCPR/C/21/Rev.1/Add.5, 8 April 1994, Comment 6.1 <https://www.refworld.org/legal/general/hrc/1994/en/26900>

which notes that it is not enough to refrain from treating similar cases differently. What is also required is a recognition that when we have cases which are truly different, different treatments are *not* discriminatory.

"To treat in an equal manner persons or groups whose situations are objectively different will constitute discrimination in effect, as will the unequal treatment of persons who's situations are objectively the same."[7]

The point of these reflections is that our understanding of what rights involve is not a one-time, static matter. Law has the potential for iterations and evolving interpretations of the very rights that may granted at one point in time.[8] As I elaborate later, it is possible that a direct bestowal of personhood to Laponia could generate both a wider range of rights and allow for greater possibilities in the development and sophistication of what rights involve and require.

In order to be able to make good on this idea, we need a richer understanding of what the personhood of Laponia might look like. That, in turn, requires a fuller account of a Sámi People's worldview, allowing us to appreciate more fully the very different structures we find in their understanding of the relationships between humans and the rest of the natural world. Only with attention paid to their worldview will it be possible to fashion rights and personhood appropriate in that particular context.[9]

Alternate Ontologies: Nature, Culture, and Animism

As a prelude to appreciating the possibilities for a Legalist reading of Laponia's imagined personhood, it is worth examining how radically different a Sámi worldview is from one commonly found in Western philosophy, political theory, and law. For starters, we might examine significant differences in how the Sámi People understand themselves and their world by way of contrast

7 UN Committee on the Elimination of Racial Discrimination (CERD), General Recommendation no. 32. The meaning and scope of special measures in the International Committee on the Elimination of All Forms [of] Racial Discrimination, CERD/C/GC/32, 24 September 2009 <https://www.refworld.org/legal/general/cerd/2009/en/70344>. Cf. "The Court has so far considered the right to [non-discrimination] violated when States treat differently persons in analogous situations without providing an objective and reasonable justification . . . [and] now considers that this is not the only facet of the [right to non-discrimination]. The right not be be discriminated against . . . is also violated when States without an objective and reasonable justification *fail to treat differently persons whose situations are significantly different.*" *Thilmmenos v. Greece*, Apply. No. 34369/97, Judgement of 6 April 2000, Article 44 <https://hudoc.echr.coe.int/eng?i=001-58561>. My emphasis.

8 This calls to mind the need for law to evolve and for it to be creative, and flexible, as discussed by Blackmun in my Introduction.

9 This demonstrates again the extent to which I agree with Tănăsescu that we must look at particular contexts to understand how RoN is deployed and developed, resisting the urge to find a universal account.

with our previous discussions of UNESCO's framing of Laponia as a World Heritage Site. In Chapter 5, we saw how Laponia qualifies as a "mixed" site, encompassing both natural and cultural values. Indeed, the very framework for evaluating sites for inclusion depends on a sharp distinction between nature and culture. Laponia is said to meet criteria for each, but the important point here is that the very method of evaluation involves a framework that makes a hard-and-fast distinction between nature and culture. Following the work of Tim Ingold, Pävi Magga, and Nurit Bird-David, among others, I suggest that it is essential to understand Sámi's *rejection* of this distinction and of associated notions. This will also entail a radically different conception of what it even is to be a person.

UNESCO's framing of Laponia as a "mixed" site seems natural from a Western perspective. Nature and culture are thought of as necessarily distinct, where culture is a byproduct of human endeavors and nature is that which is separate and distinct from humans (despite our being part of nature). To capture something unique like Laponia, we have to make use of both of these categories. But things look very different on the Sámi worldview, where nature and culture are not viewed as necessarily distinct.[10]

If indeed the Sámi People do not work with a clear distinction between nature and culture, that immediately casts UNESCO's framing of Laponia in a different light. There aren't, in short, two worlds of value—natural on the one hand, cultural on the other—which are in need or are deserving of protection. The very starting point for the legal protections offered by Sweden and UNESCO discussed in the last chapter are worlds away from how the Sámi People themselves may understand their embodiment in Laponia and elsewhere. Later, I draw on work from Tim Ingold on what an alternative conception of the relationship between nature and culture might look like. Before doing that, however, I want to dwell at some length on the animism that is said to underwrite Sámi People's worldviews. That is crucial if one of our aims is ultimately to juxtapose Laponia as understood by the Sámi People alongside our abstract Western-legal notions of personhood.

To set the stage for how the Sámi People understand themselves, their society, and what it is to be a person, it is helpful to contrast that with a Western understanding that is helpfully described by Charles Taylor in his famous essay "Atomism." There, he describes a perspective where the foundations of social contract theory, which underlie classical liberalism's view of how society should be structured, are based on a conception of humans as

10 "The Sámi do not draw a sharp distinction between culture and nature. By virtue of their different worldview, the Sámi have their own concepts for referring to the environment and nature. These often differ from those used in the dominant languages of the majority societies in which the Sámi live." Päivi Magga, 'Defining the Sámi Cultural Environment' in Sanna Valkonen, Áile Aikio, Saara Alakorva and Sigga-Marja Magga (eds), *The Sámi World* (Routledge 2022) 134, 136 <https://doi.org/10.4324/9781003025511>

autonomous and separate beings. In his terms, humans and society are best thought of as *atoms*.[11]

The image of individuals as mere "atoms" is intended to draw our attention to a vision of human beings as largely self-sufficient, where we are able to exercise our powers of rationality independently of the influence of others, forming social connections only when it is in our individual self-interest. An individual self, on this deeply entrenched view, is wholly determined by each individual's wants, beliefs, and needs, whereas societies or communities are merely transactional arrangements, with social connections as mere "contracts" entered into only when it is to our mutual advantage.

In contrast to this conception of the individual, familiar through Western metaphysis and political theory, there is an alternate view given voice to by various anthropologists. Instead of "individual," for instance, we find the helpful notion of a "*dividual*," where a person is defined not as a solitary being but instead as one who exists and is defined by reference to the particular relationships they enter into. Thus, we find in an influential paper by Bird-David a discussion of the Nayaka people of South India that makes use of this alternative notion, one that I think maps well onto the animism of Sámi worldviews I soon discuss. She writes:

> In her critically oriented comparison of the Melanesian and the Euro-American "person," Strathern argues that the irreducibility of the individual is a peculiarly modernist notion. It is not everywhere that individual is regarded as a "single entity, bounded and integrated, and set contrastingly against a natural and social backgrounds." The Melanesian "person" is a composite of relationships, a microcosm homologous to society at large. This person's objectified relationships and makes them known. She calls it a "dividual," in contrast with the [Euro-American] "*individual.*" . . . When I individuate a human being I am conscious of her "in herself" (as a single separate entity); when I *dividuate* her I am conscious of how she relates to me. . . . I am conscious of the *relatedness* with my interlocutor *as I engage with her*, attentive to what she does in relation to what I do, to how she talks and listens to me as I talk and listen to her, to what happens simultaneously and mutually to me, to her, to *us*.[12]

Here, we should note that in speaking of the relations that define a "dividual," we are including relations to other persons (which, as we'll see, may

11 Charles Taylor, 'Atomism' in *Philosophy and the Human Sciences: Philosophical Papers, 2* (Cambridge University Press 1985) 187.

12 Nurit Bird-David, 'Animism Revisited: Personhood, Environment, and Relational Epistemology' (1999) 40 *Current Anthropology* 72 <https://www.journals.uchicago.edu/doi/epdf/10.1086/200061> Used with permission from Chicago University Press.

be a much broader category than the typical Western model allows) as well as relations to the rest of the natural world. Unlike Western ontologies where we see a sharp distinction between persons and nature (and thus culture and nature), a "dividual" is comprised and defined by *all* of their relations to the world. The notion of "dividual," in other words, may be of use in helping break down that very distinction between nature and culture so deeply entrenched in Euro-American thinking.

Entwined with the preceding discussion, where we focus on the relations that define a "dividual," is an understanding where one enters into relations with a world that is not static or inanimate. Instead, we find the notion of *animism*, which is said to be a deeply important part of the Sámi People's worldview.

"Animism" as a worldview has a long history, dating back at least to Edward Tylor's writings in the 1950s.[13] Given that history and the wide range of cultures it has been applied to, I will limit myself to what I take to be most salient for Sámi. For starters, it is said that the Sámi do not have a firm distinction between spirit and matter, where, instead, all beings are considered to have souls or spirit. The different kinds of spirits are, in truth, different kinds of persons, all of which are interdependent and interrelated, yielding a rich constellation of relationships across the human and natural world.[14]

That formulation echoes what can be found in the writings of A. Irving Hallowell, who notes, speaking in this case of Ojibwa peoples, that they maintain no genuine distinction between the animate and inanimate. What matters, instead, are the *particular relations* that can occur between beings, where there seems no limit in principle to what kinds of beings are capable of social interaction.[15] To this end, Hallowell recounts a story told to him of a couple responding to the sound of thunder as one would to a social acquaintance.

> He was reacting to this sound in the same way as he would respond to a human being, whose words he did not understand. The casualness of the remark and even the trivial character of the anecdote demonstrate the psychological depth of the "social relations" with other-than-human beings that becomes explicit in the behavior of the Ojibwa as a consequence of the cognitive "set" induced by their culture.[16]

13 For a history and account of the development of the term, see Elina Helander-Renvall, 'Animism, Personhood, and the Nature of Reality: Sámi Perspectives' (2010) 46(1) *Polar Record* 44 <https://doi.org/10.1017/S0032247409990040>

14 ibid, 44.

15 Irving Hallowell, 'Ojibwa, Ontology, Behavior and World View' in S Diamond (ed), *Culture in History: Essays in Honor of Paul Radin* (Columbia University Press 1960) 52.

16 ibid, 64.

While in this example, it is thunder that is said to play a "social role," Hallowell points to another example in which even rocks can be said to stand in that relationship. " 'Since stones are grammatically animate,' I once asked an old man: 'Are all the stones we see about us here alive?' He reflected a long while and then replied, 'No! But some are.' "[17]

Similarly, Ingold dismisses the tidy distinction between the animate and inanimate:

> I do not think we can regard this [animacy] as a property that can be ascribed to objects in isolation, such that some (animate) have it and others (inanimate) do not. For life is not a principle that is separately installed inside individual organisms, and which sets them in motion upon the stage of the inanimate. . . . Indeed once we think of the world in this way, as a total movement of becoming which builds itself into the forms we see, and in which each form takes shape in continuous relation to those around it, then the distinction between the animate and the inanimate seems to dissolve. The world itself takes on the character of an organism.[18]

Pulling these various threads together, we may say that an animistic worldview is thoroughly "relational" in that what exists is understood by examining the relations that things bear to each other—where "things" is an open-textured domain that includes familiar-looking human beings, alongside non-human persons, spirits, and physical objects. The "dwelling perspective" that Ingold is offering as an alternative to one where humans are distinct from the rest of the world holds "that it is because we are, from the outset beings-in-a-world, which affords meanings to us in contexts of current action, that we can produce representations of it."[19]

Ingold gives us further help understanding this relational epistemology or metaphysics, which can again help undercut the Western-distinction between nature and culture, which drives that UNESCO framework. On his view, inhabiting or dwelling within the world makes the cultural elements as much of the encountered world as rivers or trees or glaciers.[20]

17 ibid, 54.

18 Tim Ingold, 'The Temporality of the Landscape' (1993) 25(2) *World Archaeology* 152, 164. Reprinted with permission from Taylor & Francis.

19 Helander-Renvall, *Animism*, 46. The fundamental notion of "being-in-the-world" as a primary mode of experience and knowledge, compared with a theoretical perspective which involves making abstract representations of the world, is one of Martin Heidegger's most important contributions to philosophy. Famous formulations of this can be found in Martin Heidegger, *Being and Time* (J Stambaugh tr, State University of New York Press 2010).

20 "I argue that we should adopt, in place of both these views, what I call a 'dwelling perspective,' according to which the landscape is constituted as an enduring record of—and testimony to—the lives and works of past generations who have dwelt within it, and in so doing, have left there something of themselves." Ingold, *Temporality of Landscape*, 152.

Ingold is pointing towards a mode of existence found in Sámi animism that differs radically from the ontology of traditional Western metaphysics—one that winds its way from Aristotle to the present day. What is telling in Ingold's account is that our primary experience of the world is not a detached theoretical perspective, where we stand back from our interactions with the world and others and postulate a theory that captures those experiences. Rather, it is an immersion in that world of being with others, of being alongside them in conversation, in shared activities, of hunting, of gathering, of dancing, of celebrating. It is those activities in which we are immersed where things become present for us and where we find our relations to them. We know the world through our relational interactions, our activities, in that world. And it is here that we encounter the world in its most fundamental way, a way that makes folly of the idea that, somehow, nature and culture are distinct.[21]

Personhood Implications: Difference and Tolerance

I have sought in the preceding pages to give an overview of key philosophical and anthropological aspects of a Sámi worldview through published literature.[22] I now draw out some of the legal implications, particularly as they concern the legal personhood of Laponia I've imagined. To begin, we can inquire how Laponia as a legal person fit might within the alternatives of Realist and Legalist perspectives.

From the Realist perspective, as we've seen, for Laponia to be a legal person, it would have to have one of those essential characteristics outlined in Chapter 2 and 3, say, sentience, rationality, or intrinsic value. Taking that perspective, and juxtaposing it with some of the characterizations described earlier in a Sámi worldview, we might think that it is *not*, in fact, too difficult to overlay a Realist legal conceptions onto that worldview. After all, we've seen indications that Laponia (or portions with it) might indeed be considered animate and even sentient. The collapse of the nature/culture distinction explored earlier might indicate that Sámi conceptions lend themselves straightforwardly to the Realist model: there are non-human persons that humans can relate to and communicate with. And, if we take the rejection of atomism seriously, we might imagine making the case that Laponia is itself a relational whole, a self-standing entity in its own right. And it would be an

21 I take many of Ingold's insights as grounded in the philosophy of Heidegger. Unfortunately, the latter's terminology and writing style is notoriously opaque. A helpful primer for understanding Heidegger's philosophy is Hubert Dreyfus, *Being-in-the-World: A Commentary on Heidegger's Being in Time, Division I* (MIT Press 1990).

22 Though there is a wealth of published material, there is a tremendous amount of unpublished work and untapped traditional knowledge of the Sámi People that has not yet made it into academic form.

entity that, by Sámi standards, might be deemed close enough to what the Realist is looking for to allow the Realist to welcome Laponia so conceived as a legal person. Laponia, as a legal person, would then have a rich set of rights (and possibly duties) and should be recognized as having them because it is a legal person, and it is a legal person because of its (quasi-) animate-sentient nature.

Now I expect the preceding discussion would not be particularly persuasive for those committed to the Realist model who come to these discussions from the Western-legal tradition. Simply put, I imagine many of those Realists (and others) to object that Laponia is not *truly* sentient, and it is not *really* animate. Though the Sámi People might speak roughly in those terms, continues the objection, and though perhaps we could translate and understand their animist traditions as gesturing towards a kind of sentience in some way analogous to the sentience of animals, or as active in a way that bears some resemblance to how animals are animate, those are all exceedingly charitable understandings of Laponia's nature. But when all is said and done, continues the objection, Laponia doesn't *really* have those features. Sámi may have various myths, stories, and traditions that speak of rocks and like as animate, but those are mere superstitions, and we shouldn't take seriously this proposed metaphysics of Laponia, despite the importance to Sámi culture.

By way of response to this kind of skepticism, I think we should distinguish several paths. One might say, following what I called the "assimilationist arguments" in Chapter 2, that we should consider it a point of pride for legal systems that, over time, they have both widened their recognition of the essential features of human beings and expanded the domain of beings that are recognized as having those features. We've gone from thinking that what is essential to personhood is being a property-owning white male to recognizing that that is not the essence of a human being. Perhaps it is something like rationality, or sentience, or, as discussed, a being's intrinsic value. Further, law has gradually come to recognize that there are beings who are not white males who also have those newly appreciated essential features. In that vein we might say that Laponia's personhood, premised on a continuing expansion of what is essentially human, is, in fact, permissible and would constitute another progressive step in expanding the domain of beings that are persons and that have rights. That, after all, can be seen as a driving motivation of RoN—to break through the limits of what a rights-bearer looks like.

As I've previously argued, it does indeed seem a good thing that our understanding of humanity has expanded and that, over time, we have come to widen the circle of rights-bearers. My complaint, as argued for in previous chapters, is that we are forcing would-be persons to resemble, to a sufficient degree, a paradigm, i.e. humanity, and providing generous extensions of our conception of humanity. The downside of that assimilationist approach, despite the progress it has led to, is that this reasoning still makes differences a liability. The problem from the start, we might say, was thinking that women and people of color are not legal persons because they didn't resemble the paradigm. There has been progress in revising the paradigm, no doubt. But by continuing to focus on similarities with the paradigm as the key, we perhaps blind ourselves not only to differences that might be very important but also that the differences themselves may very well be worthy of respect.

This may be because women or people of color do not share all of the values and perspectives of white males, and their differences and experiences provide a diversity that is worth respecting. That is one of the takeaways, I think, from General Comment 32 of the CERD discussed earlier. Differences can matter as much as similarities and need to be respected and treated as such. Thus, my first objection to the Realist attempt to accommodate Laponia is that it is trying to fit Laponia into a box it perhaps simply does not fit, nor, even more importantly, should it need to fit.

My second reply to the Realist gambit I've mentioned stems from the problem that arises should one attempt to argue on behalf of the Realist that Laponia truly does have the features needed to say it sufficiently resembles a human being. For as mentioned, that claim is surely to be rejected by many. Faced with an imagined Realist who wants to claim Laponia is human-like enough to qualify as a legal person and a skeptic who disagrees, how would we settle the matter? It seems unavoidable in this case that we be drawn into deep philosophical (not to mention biological) waters over the nature of humanity, of what the necessary and sufficient conditions are for something to truly be said to be animate, or sentient, or an integrated whole. And that, I suspect, is territory most jurists would prefer to not have to engage with if it can be avoided.

After all, even in the case of the animal rights movement, as discussed earlier regarding the views of Singer and Regan, the question of whether animals are *really* sentient poses a range of extraordinarily complicated questions. Are animals sentient in a way that resembles the sentience of humans? How biologically complicated must an organism be to be said to be capable of feeling pain? Does the hypothesized pain of a fish or spider matter, ethically and legally? I suspect it is these questions which have been the stumbling blocks, among others, that have hampered efforts to grant animals moral and legal rights. How much more complicated will our debates be if we find ourselves arguing over whether to accept claims from a Sámi worldview that speaks of animate rocks that can communicate with humans?

As I've said, I imagine most jurists would prefer to not be embroiled in such debates if there were another way. I think the Legalist view does offer an alternative, where we seek shelter in the creative opportunities it seems to offer. A crucial upshot of Chapter 4, after all, was that we need *not* tightly link legal personhood to features deemed essential to human beings. Law has the potential for an expanded understanding of what legal persons are and of what the domain of beings is deserving of the full menu of rights and duties that are already bestowed on some of the natural world, i.e. human beings. As the example of corporations shows, we already extend the domain of legal persons beyond the human person. Why not for Laponia? That's the point of being a Legalist strategy in the first place: there are no, or perhaps only minimal constraints, on what can be a legal person.

But while that may make for a nice slogan, I concede that it does not yet count as an *argument* for Legalism. It is simply a restatement of my by-now-familiar claim that Legalism does not have (many) *a priori* constraints on who or what can be a legal person. What I think does help advance the argument, though, is the *tolerance* that Legalism may allow for. For if the heart of Legalism is that, in theory (nearly), anything can be a legal person, at one level, this is a plea for *metaphysical or ontological* tolerance. There are different kinds of persons, some of which do and some of which don't map onto Western conceptions of the human being. Legalism in that way opens the door for various conceptions of persons, without requiring us to take a stand on which worldviews are true and which are mere superstitions. In the present case, the open-ended nature of Legalism allows us to take at face value the distinct metaphysics and epistemology we find in Sámi animism and, thus, to avoid becoming overly embroiled in the kind of metaphysical and ethical debates many would seek to avoid.

I think, however, there is a deeper and more important kind of tolerance possible here, one that fits more squarely within a certain perspective touted as an important virtue for Western liberal democracies. There are many ways, of course, to characterize liberalism, and an investigation into them would take us very far afield. A particularly powerful version, though, can be found in the writings of John Rawls. Simply put, what distinguishes political liberalism from older and alternative conceptions for grounding political structures is that liberalism rejects the view that there is a common conception of "the good." That is not one particular vision of what the good life involves, of what individuals should value, or what the specific substantial goals of a society should be.

> Political liberalism was born out of a crisis in this principle of legitimacy. The ambition underlying all versions of liberalism has long been to define the common good of political association in terms of a minimal moral conception—that is, a basic value or set of values that most citizens share despite even their many important differences. Political principles are neutral—and thus satisfy the liberal principle of legitimacy—insofar as they can be justified by reference to such shared values, *without assuming the validity or truth of any particular (controversial) conception of the good life.*[23]

23 John Rawls, *Political Liberalism* (Columbia University Press 1993) 3. My emphasis.

My suggestion here is that Legalism allows for not only a metaphysical tolerance but also allows for a value/political tolerance. We can accept a Sámi People's worldview, which sees the world differently both in its metaphysical structures and in the values that should guide one's life, by allowing a diversity of views. We need not wrestle with which view is correct, and in turn, we allow ourselves a more open, diverse, and tolerant society that acknowledges difference—including, in particular, who the beings are in a legal system that are deserving of rights. The creative uses of law that Legalism offers widen the scope of what can be a legal person and, in its wake, opens the possibility for diverse values to sit more easily side by side. It perhaps allows us to use law in pursuit of fostering and facilitating respect for difference. As part of a political theory that values and respects difference, Legalism seems well suited to that goal.

Personhood Implications: Persons and Rights

I argued earlier that direct bestowals of legal personhood are the most powerful way of making something legally visible, carrying with it the potential for a broader range of rights that can evolve than might be done by legal stipulations that a body of water, say, has a defined set of particular rights. I wish to expand upon this now, in an effort to show the further potentials that might follow in granting Laponia legal personhood in what I called the direct manner.

Here, an important distinction should be made. On the Legalist understanding of personhood I've been developing, there are no requirements that the concept of a legal person track features deemed essential to human beings. In that way, Legalism can be seen as neutral on the nature of legal persons. And as I've suggested, this neutrality allows to understand the broad range of things that law has deemed legal persons, from infants to businesses, from States to international organizations, and of course, the various rivers, ecosystems, and the like that are part of RoN.

Still, in another sense, I think there is going to be a tight connection between the nature of the legal person in question and the kinds of rights (and duties) appropriate to it. For alongside the general, abstract point that being designated a legal person opens a wider range of possible rights than might be found by first focusing on a set of clearly demarcated and articulated rights, at the same time, the nature of the person in question is going to structure or influence the kind of rights it can be accorded. Different persons, we might say, provide for different rights, where again, the *nature of the person in question* is going to drive which rights that person has.

Two brief examples can help make this point, which I then use to illustrate the potential for rights that might be said to flow from the personhood of Laponia. First, in the case of corporations in the United States, which have long said to be legal persons, the particular rights that flow from that personhood reflect the nature of it being a corporation, as opposed to, say, it being an infant. Of course, and importantly, as I've remarked, that set of rights is not fixed and is subject to development, evolution, and ongoing re-interpretations. Still, we might say that the rights that corporations have are indicative of the kind of thing a corporation is. It seems strange to say that a non-sentient person, such as a corporation, could have rights against being tortured. It just isn't the kind of thing that needs protection against harms to its sentience since it isn't sentient.[24] The kinds of rights that corporations have been given and their evolution over time is helpful in letting us understand that the particular rights a person has is, in some way, going to be tied to the nature of that person, even if those rights are subject to expansion and development. For instance, the US Supreme Court granted corporations rights against warrantless searches in 1978; in 2010, it provided free speech rights to corporations, rights that prohibit the government from restricting independent political expenditures; and in 2014, they were granted rights to free exercise of religion.[25]

The point, again, is that while personhood allows for a wider range of rights to be granted that person (compared with a stipulation of just a few rights, as we've seen the examples from Colombia and Ecuador), I maintain that the rights that are given to a legal person (both at a time and over time) are ones that are judged to be appropriate to the nature of the person in question. As noted earlier, rights against warrantless searches were *eventually* decided to be appropriate for corporations in a way that were not initially. So, too, for free speech rights and how those speech rights were to be understood, including that monetary contributions to political campaigns count as expressions of free speech. A human being, in contrast, as a legal person had those kinds of rights built in from the very start, at least in theory. So, too, for the exercise of religious rights, which only over time were deemed appropriate for corporations, lagging far behind their being granted to human beings. Put differently, if all legal persons have the same range of rights, then corporations would have had those speech and religious rights from the start.

By way of a second example, briefly consider again our discussion of whether AI engines can or should be deemed to be legal persons. Without wrestling with any of the details of that debate, we can at least recognize that were AI engines designated as legal persons because they are allegedly intelligent or sentient, we will have to structure the kinds of rights *appropriate for them* given that the kind of intelligence/sentience they may have is likely to be

24 Were certain animals given personhood and rights, it would be absurd to imagine them having the right to a university education. They are not the kind of beings that "fit" with that kind of right.

25 Hon. Katherine B Forrest (Fmr.), 'The Ethics and Challenges of Legal Personhood for AI' (2023) 133 *The Yale Law Journal* 1196.

radically different from the sentience of human beings and perhaps animals. For example, if a particular AI system is not located in any one particular place but perhaps spread out on different computers in different countries, we will have to tailor rights to, say, communication, or speech, to fit their multiple locations, among other complexities.[26]

The point, again, is that the different natures of various legal persons is going to determine different sets of rights that are appropriate for the person in question, both at a given time and how those rights might evolve and be understood over time. Thus, to revert to a heuristic from Chapter 1, we may find it useful to think of particular persons with particular rights, where the emphasis now is on how the nature of that person may structure or drive the various rights that person has.

Person$_{\text{Human Being}}$: Set of Rights A
Person$_{\text{Corporation}}$: Set of Rights B
Person$_{\text{AI}}$: Set of Rights C

If these abstract reflections are on point, then it would follow that there will be a range of rights appropriate to Laponia's personhood, given that Laponia is a radically different kind of person than a corporation, or a human being, or an AI engine.

Person$_{\text{Laponia}}$: Set of Rights D

As an example of what might be possible and/or appropriate, recall UNESCO's characterization of Laponia's "natural heritage," which mentions both the presence of glaciers and unique mire complexes.

> The nominated area contains all the processes associated with glacial activity such as U-shaped valleys, moraines, talus slopes, drumlins, presence of large erratics and rapidly flowing glacial streams. It has excellent examples of ice and frost action in a tundra setting including formation of polygons and an area of spectacularly collapsing and growing palsa mounds. Glacial rivers originating in the snowfields continue to cut through bedrock. Large unvegetated areas illustrate the phenomenon of weathering. . . . The vast mire complex of Sjávnja/Sjaunja is the largest in Europe outside Russia. This area is virtually impenetrable by human beings except during winter. The Laponian area has primeval coniferous forest with dating indicating ages as old as 700 years.[27]

26 ibid, 1177. The discussion there covers some of the challenges involved in making AI legal persons, as well as the kinds of creative use of laws, rights, and duties that might be needed.
27 'Laponian Area' <https://whc.unesco.org/en/list/774/>

Taking our point of departure from complex developments in the rights of rivers, perhaps there could be *glacier rights*, or *mire rights*.[28] Laponia's present status as a protected area could be strengthened, as I've suggested, by both giving it rights of its own as well as even personhood. If one of the motivating goals here is protection of the environment, then the kinds of rights intended and required for that protection should seemingly be suited to the distinctive features of the person in question. Laponia, as a unique mire complex, along with its ongoing glacial activity, would seem to call for a different set of rights than those intended to protect a river or a forest.

Engaging in creative reflections about the unique nature of Laponia could usher in not only rights tailored to the kind of person it would be but would also bring along a host of other considerations that again would be appropriate for it and its unique nature. This might include revising and strengthening the guardianship arrangements for enforcing and protecting those rights, or perhaps specified remedies when the interests of Laponia are harmed, guided by ongoing scientific knowledge of what is needed for mires, for instance.

Going a bit further, we might attend to other features of Laponia that make it unique, which again would seem to open space for its own distinctive set of rights. Were we to take seriously some of the Sámi worldview elements discussed and view Laponia as, in some way, a unified, relational whole, one that is, in some way, sentient, without a clear distinction between its animate and inanimate aspects, we again could find ourselves pointed towards the need for rights appropriate to it in its particularities. What's more, the freedom of Legalism seemingly opens the door for radically different kinds of persons that, on the present argument, point towards a wide range of very different sets of rights.

Finally, as discussed, Laponia was originally designated by UNESCO not only for its natural heritage but also for its cultural elements. And while our discussions have perhaps cast doubt on that distinction, Laponia remains an ongoing lived-in area, where the Sámi People continue the age-old practice of reindeer herding. Creative uses of law can be imagined where Sámi rights to self-governance and self-determination are further enhanced and developed, giving Sámi perhaps greater control over hunting and fishing rights in Laponia. Importantly, because Laponia is an area that Sámi depend on, the construction of rights for it will not look like the kinds of rights we might imagine for untouched wilderness regions. A balance no doubt would have to be found that respects the interests of all the relevant parties—from the mires and glaciers to the reindeer and Sámi traditions and practices—though

28 For an extensive summary of the jurisprudence on river rights that have emerged in recent decades, see Cristy Clark, Nia Emmanouil, John Page and Alessandro Pelizzon, 'Can You Hear the Rivers Sing? Legal Personhood, Ontology, and the Nitty-Gritty of Governance' (2020) 45 *Ecology Law Quarterly* 787 <https://doi.org/10.15779/Z388S4JP7M>

finding the appropriate balance is part of any legal regime that bestows rights on a personhood.

Much of this is admittedly vague. Still, I think the important points still stand. As we've seen in our discussions of rivers in Colombia, Ecuador, and New Zealand, the resources of law have emerged in particular contexts—social, political, and economic—which have, in turn, shaped the form that RoN have taken in those contexts. Laponia would be no different. What I hope to have added is a recognition that we would expect a certain kind of fit, for lack of a better word, between the nature of the legal person in question and the exact form the rights appropriate to that person take. The open-ended nature of rights that come with something's being a legal person does not mean that all rights are going to be appropriate for every legal person. Our discussions of the unique features of Laponia have given us some insight into what kind of entity it would be were it given personhood. And from that, we can imagine that resources and creative uses of law could be used to implement a picture of Laponia as a legal person with, again, a set of rights that provide for an appropriate fit. What exact form that might take is beyond what I'm in a position to articulate. But I hope that this all points in an optimistic direction.

References

Berlin I, 'Two Concepts of Liberty' in H Hardy (ed), *Liberty* (Oxford University Press 2002).

Bird-David N, 'Animism Revisited: Personhood, Environment, and Relational Epistemology' (1999) 40 *Current Anthropology* 72 <https://www.journals.uchicago.edu/doi/epdf/10.1086/200061>

Clark C, Emmanouil N, Page J and Pelizzon A, 'Can You Hear the Rivers Sing? Legal Personhood, Ontology, and the Nitty-Gritty of Governance' (2020) 45 *Ecology Law Quarterly* 787 <https://doi.org/10.15779/Z388S4JP7M>

Dreyfus H, *Being-in-the-World: A Commentary on Heidegger's Being in Time, Division I* (MIT Press 1990).

Forrest KB, 'The Ethics and Challenges of Legal Personhood for AI' (2023) 133 *The Yale Law Journal* 1196.

Hallowell I, 'Ojibwa, Ontology, Behavior and World View' in S Diamond (ed), *Culture in History: Essays in Honor of Paul Radin* (Columbia University Press 1960) 52.

Heidegger M, *Being and Time* (J Stambaugh tr, State University of New York Press 2010).

Helander-Renvall E, 'Animism, Personhood, and the Nature of Reality: Sámi Perspectives' (2010) 46(1) *Polar Record* 44 <https://doi.org/10.1017/S0032247409990040>

Ingold I, 'The Temporality of the Landscape' (1993) 25(2) *World Archaeology* 152.

International Covenant on Civil and Political Rights (adopted 19 December 1966, entered into force 23 March 1976) 999 UNTS 171 art 31(1).

'Laponian Area' <https://whc.unesco.org/en/list/774/>

Magga P, 'Defining the Sámi Cultural Environment' in S Valkonen, Á Aikio, S Alakorva and S-M Magga (eds), *The Sámi World* (Routledge 2022) 134 <https://doi.org/10.4324/9781003025511>

Rawls J, *Political Liberalism* (Columbia University Press 1993).
Stone CD, *Should Trees Have Standing? Law, Morality, and the Environment* (3rd edn, Oxford University Press 2010).
Taylor C, 'Atomism' in *Philosophy and the Human Sciences: Philosophical Papers, 2* (Cambridge University Press 1985).
Thilmmenos v Greece, Apply. No. 34369/97, Judgement of 6 April 2000, Article 44 <https://hudoc.echr.coe.int/eng?i=001-58561>
UN Committee on the Elimination of Racial Discrimination (CERD), General Recommendation no. 32, The Meaning and Scope of Special Measures in the International Convention on the Elimination of All Forms [of] Racial Discrimination, CERD/C/GC/32, 24 September 2009 <https://www.refworld.org/legal/general/cerd/2009/en/70344>
UN Human Rights Committee (HRC), CCPR General Comment No. 23: Article 27 (Rights of Minorities), CCPR/C/21/Rev.1/Add.5, 8 April 1994, Comment 6.1 <https://www.refworld.org/legal/general/hrc/1994/en/26900>

Chapter 7

Concluding Strong and Weak Legalisms

I set out in this book to examine how competing conceptions of personhood might shape the RoN movement. In particular, I've focused on the notion that legal persons need not resemble human beings. Or, put a bit more precisely, that there are not *a priori* constraints on legal personhood that require it to track claims about the essential nature of human beings. This Legalist understanding of personhood, I've argued, makes for a better fit for RoN. Nature need not be rational, nor sentient, nor have a will, nor have any other feature we find in humans for nature to be the kind of thing that can have rights of its own and to even be a legal person.

It is this open-ended nature of legal persons which gives Legalism its power. Among other virtues explored previously, Legalism additionally facilitates more easily than Realism a key desideratum for law, namely that it be an ongoing, evolving enterprise. Without trying to offer a full account of what law is, we can safely say that, if nothing else, law provides a means for society to wrestle with its political, social, and economic needs. If nothing else, it is an important tool for helping a society reach its goals, both long and short term. If law has at least this instrumental value, then decisions about which legal devices to employ and how to understand them turns, in part, on how well those legal devices help facilitate or achieve those societal ends and goals.

The invention and development of corporate personhood is a case in point of how a society can create legal tools to facilitate, *inter alia*, various economic goals. So, too, we've taken a brief look at various, sometimes competing, goals that drive debates over the nature and uses of AI, including whether some of society's interests can (best) be met by granting AI ownership rights and possibly even personhood. These highly abstract reflections on the nature of personhood have then been employed with an eye to which account of personhood is best (or at least better) suited to the overriding problems of climate change and the ongoing destruction of our environment, along with queries on the possibility of novel rights-regimes being employed as one of various legal bulwarks.

DOI: 10.4324/9781003740186-8

So we find ourselves asking whether there are new and novel legal devices that can help strengthen the protections of nature. If there are competing ways of achieving those ends, then perhaps the question becomes which of those devices provides the flexibility and the potential for development and evolution that will be needed as society wrestles with ongoing climate destruction. What legal tools and what construal of those tools best assists in that larger project? That is one important way to understand the debate between a Realist and Legalist understanding of personhood. I've sought to argue that a Legalist understanding of personhood positions it to better achieve the goals expressed in RoN.

A key moment in this book is the argument that as we contemplate the possibility of granting personhood to nature, without requiring the non-human parts of nature to resemble the human parts, we recognize that nature can have interests and interests that need legal protection in various forms, without requiring those interests to all be grounded in sentience. That was the upshot of Chapter 4, where I argued that, with corporations, we already recognize the existence of legal persons that are not sentient. If we can do it for corporations, then, by parity of reasoning, we should be able to consider doing it for nature. If, that is, the need is there and if bestowal of rights and personhood does further the cause.[1]

Similarly, and more broadly, I've suggested along the way that a Realist understanding of personhood fares worse than a Legalist reading in that the Realist seems to allow the non-human parts of nature to count as legal persons only to the extent that the non-human parts of nature resemble human beings. In so doing, we are asked to engage in various anthropomorphic twists that do not stand up to critical scrutiny. In contrast, as I've suggested, the flexibility of Legalism allows us to appreciate differences and not to be solely concerned with similarity.

Limits on Personhood?

I've stressed throughout what I take to be the key merit of Legalism, namely, the freedom it gives for finding and creating legal persons that are not bound to conceptions of humanity, however we choose to define the latter. The open-ended character of Legalism for adopting new and novel legal persons is well-suited to RoN and better suited than Realist renderings. Nature (the realm that extends beyond humans, that is) need not be rational nor sentient in order to be a legal person. I've suggested as well that we also need not be constrained in grounding personhood in such unwieldy notions as intrinsic value or inherent worth.

But in the face of this open-ended view, one might reasonably wonder if there are any limits to what can be a legal person. After all, if we allow

1 Or, better put, if it helps achieve one of the various goals motivating RoN.

ecosystems, rivers, and glaciers to have their sets of legal rights and to be legal persons, is there anything in principle that cannot be granted the status of a legal person, a legal entity, a subject of law? A skeptic to all of this may be tempted to ask both rhetorically and mockingly: So can the number 7 be a legal person? The color blue? AI systems that we discussed in Chapter 4's Appendix? Put non-rhetorically, one may wonder whether there are indeed any limits, with the implication that conceding there are none amounts to a *reductio ad absurdum* of my theory.

Now at one level, the skeptic can be seen as wondering (perhaps only half-seriously) about those alleged examples. At a deeper level, I think the skeptic desires an understanding of the full contours of the Legalist understanding of personhood. That is a fair request. It is reasonable, after all, to inquire into the boundaries between being a legal subject and a mere object: to understand which uses of personhood are legitimate and which are not. The distinction between Strong and Weak Legalism arises at least in part from this sort of problem.

Strong Legalism, we may say, sets the bar as low as possible, with no a priori constraints on being a legal person. It is this view which we encountered in Chapter 1 where it has been claimed in essence that legal personhood is merely an empty slot, one that can be filled in with anything. Not only are there no requirements that legal persons resemble in any human beings, the position of Strong Legalism has no restrictions at all. Weak Legalism, on the other hand, as we've seen, does have one a priori constraint, one threshold as it were that something has to cross to be legal person. Namely, having interests is a necessary condition for personhood. Crucially, as I've said repeatedly, those interests need not resemble in any manner the interests of humans, where, traditionally, that has come in the form of a being's sentience.

With that in mind, one strategy for responding to our skeptic is through a case-by-case examination of putative counter-examples. That was the strategy of the Chapter 4 Appendix discussion of AI systems and their claims for personhood. It was suggested there that AI systems don't have interests and, as such, are not capable of having rights. Without rights, so, too, we lack grounds for personhood. On the other hand, I've argued that nature, including non-sentient parts, such as ecosystems, *do* have interests and, thus, clear the (admittedly low) bar for being a rights-bearer, thereby qualifying as a legal person.

Thus, when confronted with a putative counter-example, one first asks if it is the kind of thing that is capable of having rights.[2] That, in turn, requires it being the kind of thing that has interests. So the first substantive question one must ask is whether the thing in question does have interests. As I've

2 The echo of Raz's initial question, of whether a thing is the kind of thing capable of having rights is intentional.

explained in Chapter 4, to have an interest is to have a stake in a condition a change to which is better or worse for a thing. That thing, again, must be subject to changes that can make it better or worse off.

To return to the two alleged counter-examples, 7 and blue, we should note first that, arguably, neither 7 nor blue are the kinds of things that can change in the first place. Blue cannot change into purple, though, of course, a blue thing can. The blue itself might be replaced by another color but isn't the kind of thing that can survive a change into a different color. So, too, for 7, which can't undergo a change that makes it 8. 7 bananas can become 4 if I'm sufficiently hungry, but the number 7 cannot become 4.[3]

So at one level, we can dismiss at least some of those putative counter-examples put forth ostensibly as a *reductio ad absurdum* on the Legalist position. But the deeper issue that the skeptic is perhaps driving at remains. On the Realist view, as we've seen, the concept of personhood is fairly determinate. That isn't to say it hasn't and cannot change, for in focusing on rationality, say, or sentience, or even intrinsic value, the Realist is pointing to different criteria for personhood. Still, those changes are within a fairly limited domain, as they are simply different accounts of the essence of a human being. It is the (relative) determinateness of our notion of a human being which the Realist maps their accounts of legal personhood onto.

In contrast, the Legalist account of personhood seems quite indeterminate. If to be a legal person is just to be the bearer of right and duties, and the intrinsic nature of the subject of those rights and duties is not important, it can seem hopelessly unmoored. Personhood is simply not tied to any particular concept of a human being, nor it would seem to any particular subject at all. Thus, what the Legalist sees as a strength in this way of thinking about legal subjects, a skeptic might find impossibly vague or indeterminate.

There are at least two sorts of replies available here. First, we can recall that this project has tried to determine the *relative* merits of a Legalist view compared with the Realist, for an account of personhood that can be of benefit for RoN. I've tried to argue that as far as RoN goes, a Legalist account of personhood does make for a better fit. The Legalist is not bound to characterizing (non-human aspects) of nature in anthropomorphic terms. The Legalist is not forced to take a stand on which entities have intrinsic value. The Legalist conception of interests, following the example of corporations, recognizes that interests don't require sentience. Nor is the Legalist bound to what I've called assimilationist arguments. There is not only one model for being a legal person. Unlike our skeptic who may see that as disquieting, the Legalist embraces the open possibilities it implies for developing and generating new rights for nature. The personhood appropriate for nature might have some overlap with human beings *qua* legal persons, and/or with the personhood

3 The numeral "6," which represents the number 6, can, of course, change. But the number can't.

appropriate to NGOs, or States, or small businesses, or strong AI models that might appear one day, or with the personhood of whales, should they ever be given legal personhood. Concepts evolve, and their developments often arise as modifications of present and also older versions, often with changing metaphors and analogies that are used to flesh out the concept.

Our concept of an atom, for instance, has evolved from its initial, literal meaning of something indivisible, to what was famously known as the "plum-pudding model," where negative and positive charges were bound together into a single mass, changing again with the help of solar system model, with electrons circling a nucleus, before changing again with talk of "shells" and quantum levels, with electrons being characterized by properties also built on analogies, such as "spin."[4] The point being that just as our concept of "person" has already undergone changes, beginning with the notion of persona as a theatrical mask, as discussed in Chapter 1, so, too, can we expect it to change and evolve. Both the meaning (intension) of the concept "person" and its extension (the set of things that are persons) should be expected to change. Ongoing debates about the nature of personhood are themselves part of the process that will determine which changes occur, which might be temporary and which might be more permanent.

Wittgenstein offers a nice image of the evolution and development of language, comparing it to an ancient city.

> [A]sk yourself whether our language is complete;—whether it was so before the symbolism of chemistry and the notation of the infinitesimal calculus were incorporated in it; for these are, so to speak, suburbs of our language. (And how many houses or streets does it take before a town begins to be a town?) Our language can be seen as an ancient city: a maze of little streets and squares, of old and new houses, and of houses with additions from various periods; and this surrounded by a multitude of new boroughs with straight regular streets and uniform houses.[5]

The second reply to the skeptic, in addition to pointing out the relative merits of Legalism versus Realism, addresses head-on both the potential, and just as importantly, the limits of legal theorizing. As a part of nature, humans have generated a legal system that grants a special status to certain portion of that natural world—namely, ourselves. And as we need to, we extend, shrink, modify, and develop the structures of that special status, in part by deploying notions of legal personhood, rights, and duties, among others. Crucially, that is, law is not static. It responds to the complexities of economic, social,

4 'The History of the Atom—Theories and Models' <https://www.compoundchem.com/2016/10/13/atomicmodels/>

5 L Wittgenstein, *Philosophical Investigations* (Basi Blackwell 1953) §18.

and political forces, not to mention sociological and philosophical ones and not to mention our understandings of the world from science. I think the dynamic nature of law is a good-making feature. Law that is unresponsive to changing conditions and needs would be ossified and ultimately of little use.

The flip side of this dynamic nature of law, however, is that we cannot fully spell out in advance what the limits are (or should be) for such foundational legal notions as rights and personhood. The flexibility of law we desire and need, I believe, has as a consequence that perhaps the skeptic is asking too much of theory of law in wanting to know here and now what the limits of personhood are. We can't screw up our legal eyes to see the future. While we can try to anticipate future needs and changes, we are not in position to know what future legal needs will present themselves to us, nor can we know here and now what will be required to address those needs.

Nor in that way are we able to discern all the possibilities that are built into the concept of *personhood* and tease them out, the way we can analyze the concept of "bachelor" and get "unmarried male." If law is a living enterprise one created by humans to, among other things, assist in their interactions with others and rest of the natural world, then it will be those forces which shape where and how personhood is defined.[6] What are our needs? What are the projects our society has decided upon, and how effective can applying rights and personhood to rivers and glaciers be to the projects we as a society have articulated, explicitly and otherwise? How we answer those questions will determine what we decide has rights, what kinds of rights they might be, and ultimately, who is a legal person.

As I see it, instead of seeking the purity and limits of the concepts of rights and persons as the skeptic seems to want, perhaps the focus instead needs to be on effectiveness of legal instruments to achieve the goals societies set from themselves. Which instruments are better in achieving those goals? If we decide that a pressing goal is protection of the environment, we must inquire into which tools are best positioned to achieve those ends, including tools that recast the nature of legal personhood and the rights that are appropriate to those newly conceived legal persons. Ultimately, I believe a Legalist strategy with its open-ended structure is better suited to these tasks.

Weak Legalism and the Importance of Interests

To round out this discussion, we should return to the distinction between Strong and Weak Legalism, the former with seemingly no limits on personhood,

6 Continuing the analogy with conceptions of the atom, we are, and should be, open to new discoveries and newly cast analogies for our understanding of fundamental building blocks of reality. Science, like law, is an ongoing endeavor, one that must be responsive to both discoveries in the empirical realm as well as conceptual revolutions.

while the latter requires persons have interests, though, again, those interests need not be sentience-based. As we saw in considering some of those imagined counter-examples, the mere presence of a limiting feature—one that puts a necessary condition on being a person in and of itself—is going to narrow the domain/extension of legal persons, compared with the Strong Legalist view, whose extension seems limitless.

Weak Legalism, in virtue of having a single constraint, as opposed to none for Strong Legalism, is going to rule out some things as non-persons that Strong would allow. That's just a point about the difference between "weak" and "strong" in general and has nothing to do with the particular constraint I've focused on which characterizes Weak Legalism, namely, the requirement that legal persons have interests. But why interests? Why is that the feature I focus on? Why is that the one I choose to put to work on behalf of RoN? Why not "legal persons must have a connection with city of Vienna"? Or they must somehow be connected with the color purple? Those criteria would also shrink the domain of persons, compared with Strong Legalism. So why interests? And crucially, to return to RoN, how is focusing on personhood with its focus now on interests, and not sentience, rationality, autonomy, or will, among others the best way to help ground a theoretical project of giving rights to nature, nature now as legal person?

Thinking back to Stone's groundbreaking essay from 1972, we were asked to imagine whether trees could have legal standing, where that standing would, as the saying goes, get trees (and other parts of nature) in the proverbial legal door. Were trees to have legal standing, they would be able to participate in, if not actually initiate, legal proceedings to address harms they suffered. And that is the key; for if there is to be standing, there must be harms, either actual or imminent. Of course, something cannot suffer a legal harm unless there is something in or about that thing that makes it vulnerable to harm—that it has something that can be harmed. That is where interests enter the picture. To change metaphors, interests are the lowest common denominator across the broad range of legal entities that have experienced something across the broad range of harms that can be inflicted—whether it is physical injury, loss of employment, violation of a Constitutional right, breach of contract, defamation, medical malpractice, etc. Interests are the lowest common denominator—not in the sense of being lowly *per se* but rather of being the most fundamental possession something can have that is recognized (and protected) in a legal system.[7]

7 We've seen earlier Raz's account of rights and the connection with interests and is worth pointing out again how deeply entwined those notions are. "To say that a person has a right is to say that an interest of his is sufficient ground for holding another to be subject to a duty." Joseph Raz, 'Legal Rights' (1984) 4 *Oxford Journal of Legal Studies* 1.

Because interests are so fundamental—in being necessary for legal standing, in being valuable enough to be legally protected, and in reaching all the way across the legal spectrum, from corporate and criminal law to administrative, property, copyright, and international law—it has been essential to determine what the essence of interests is in the first place and then to inquire what kinds of entities do, and which don't, have interests. Weak Legalism, as I've used the term, tells us that a necessary condition for rights and, thus, for the legal personhood of a thing that comes along with having rights is that the thing in question has interests. That is the rock bottom, most basic feature a thing can have that pulls it within the legal realm. Not sentience, not rationality, not autonomy, not a soul. Thus, in my quest to articulate the conditions of legal personhood, I've relied heavily on that notion. And I've, in turn, argued that nature does, in fact, have interests. Thus, not only would Legalism be preferable to Realism if we were simply wondering which would be a preferable way of understanding right and personhood if nature had rights and personhood, but also, we've gone on to see that, because nature does have interests, it is indeed deserving of rights and personhood. In a word, nature has what it takes.

The American philosopher David Lewis famously argued for what he called "modal realism," whereby possible worlds, a staple of contemporary metaphysics, were real, as real as this world. He noted that when he presented his view, he was often met with an incredulous stare, as people found his theory too strange to believe. As he noted, though, he didn't know how to refute an incredulous stare. There are only arguments that can be given. In our discussions of interests, we've seen theorists, such as Raz, Feinberg, and Kurki, among others, grant that interests are needed for something to be a person but they've gone on to add other crucial elements, such as sentience, or even further, that the beings with interests must also be of ultimate value. This was done to rule out having to concede that things like houses and artworks should be considered legal persons. I, on the other hand, have opened the door to granting personhood to the strangest of entities, which surely would generate its own round of incredulous stares. Like Lewis, I don't have an argument against incredulous stares, and the strangeness of a theory is not on its own grounds for rejection.

On the Weak Legalism I have defended for giving nature personhood, having interests is necessary for personhood. I don't, however, want to say it is sufficient. But the reason is not, contrary to Raz, Kurki, and others, that I think there is some additional intrinsic feature needed alongside interests or that the interests have to take a certain form by being, for instance by being the interests of sentient beings with ultimate value. First, as I noted in the Introduction, my discussion of legal visibility, at the lowest level, law does have a way of protecting entities it finds worthy of protection, without giving those entities rights in and of themselves nor by granting them personhood. We've explored UNESCO's wide range of place and things that are legally protected but which do not have the protection afforded to those things that have rights or that are legal persons. So something's having interests might be sufficient for it to be given the lowest level of visibility. Even those who object to animals having rights or being persons will still grant the appropriateness of giving them some kind of protection against wanton cruelty and torture.

What really drives the move from "*a* has interests" to "*a* has rights" and "*a* is a legal person" is less about the properties of that thing, *a*, and more about the social-economic-political system in which law is embedded. Chopra and White put it nicely in their discussion of whether to assign personhood to AI systems: "Considering artificial agents as legal persons is, by and large, a matter of decision rather than discovery, for the best argument for denying or granting artificial agents legal personality will be pragmatic rather than conceptual."[8]

Early on in this study, I mentioned a point from Anna Grear that is worth revisiting. She suggests we use the term "legal entity" instead of "legal person," which she finds useful for a number of reasons, the most germane at present is the claim that it draws attention to how the whole enterprise involves something artificial—in that these are categories we invent and deploy. We don't discover rights and legal persons lying about with the furniture. We create them.

Our use of the term "person," particularly when discussing so-called "natural persons," can encourage us to think that there is a natural, even necessary, linkage between the notion of "legal person" and person in the biological-metaphysical-religious sense. That, in turn, encourages us to think of legal persons as of necessity and in the most basic sense, as about humans, and that law itself is and must be primarily about humans. That can blind us to the creative uses of law that are possible, which can include the possibility of extending the range of law's persons, law's entities, to a much wider realm of beings. We must keep in mind, again, that all legal persons are themselves created and are functions of a legal system that has created a certain class of beings—call them persons or law's entities—in order to serve social-economic-political purposes.

There is no class of persons more natural than another, for they are *all* constructed. Keeping that in mind, finally, can make it easier to accept that if there are genuine social needs for making nature one of those legal entities, we would be as justified in granting that status as we would be in making corporations legal persons or of making mentally competent adult humans legal persons, for that matter. Nature, as a legal person, is no more and no less artificial than any other legal person; they are all artificial in the sense of being constructed to serve our various social needs. To change metaphors

8 Samir Chopra and Laurence F White, *A Legal Theory for Autonomous Artificial Agents* (University of Michigan Press 2011) 154.

and return to the insight of Chopra and White, we err in thinking the proper method in law is to cleave "nature at its joints," in this case, a metaphysically real distinction between subject and object, or person and non-person. There is no discovery at all, much less more or less accurate ones. There is simply stipulation, or as they put it, "decision."

All that said, still, there are many who will continue to balk at the idea of things like nature having interests. Even if I've been able to alleviate some worries that, in granting that nature has interests, I'm doomed to giving the most absurd things of having interests as well, say, 7 and blue, I don't doubt that many will find themselves with the nagging feeling that mere physical objects cannot have interests and thus can't have rights and be persons. And that my letting them in, as it were, given my low-bar for personhood, reveals a fundamental flaw in the entire system.

In response, I've tried to take the sting out of having such a low bar for having rights and being a person by shrinking the distance, as it were, between the things we find it easy to give personhood to (humans) and things we find it challenging to (glaciers, ecosystems, trees, etc.) by pointing out that, following Grear and others, that all persons are creations of the law.

Still, some will simply refuse to accept that microwaves and tabletops can have interests, even if I've addressed their worry that, suddenly, we'll be surrounded by too many persons. Interests are not sufficient, I've claimed, for there must be a genuine social need at hand that is served by the creation of a new person. But because I'm loath to assert in advance, a priori, what the limits will be to those social needs, I'm not in a position to rule out that this microwave, like this ecosystem, might one day need to be granted personhood. But a natural way to cut off that line of argument is to just deny that things like microwaves have interests. So let us look more closely at the claim.

I do think it is in the interests of the painting of Mona Lisa to be kept out of direct sunlight, just as it is in its interests to be behind protective glass in case someone is determined to spill oil or spray paint on it. Mona Lisa's fading from exposure to light makes that painting worse off. Compare that example with the idea that it is not in the interests of a Mies van der Rohe building to have its glass windows covered by shutters.[9] Seeing shutters put up on it would trigger in me a similar kind of revulsion in seeing the Mona Lisa destroyed by vandals who poured orange paint on it. But are the cases the same?

In the Mies case, one might be inclined to say that what is worsened is my aesthetic experience.[10] It is not in *my* interest for the building to have shutters installed. That is importantly different from saying the building itself is made worse. In the Mona Lisa case, my aesthetic experience is

9 Thanks to Mary MacLeod for the example.

10 Not to mention the market value of a shuttered Mies building.

also made worse. And so it is also not in my interest to have it spoiled that way. But in the Mona Lisa case, I find it easier to say as well that the painting is made worse. Perhaps it is because, in that case, we are talking about the destruction of the painting, which seems an obvious way in which it is worse off. Shutters on the Mies building, in contrast, might be bad for me, but as the building hasn't been destroyed, perhaps that is why I'm less inclined to say that the building has been made worse off. If something analogous to the painting were to happen to the building—say, a cantilevered section collapses after an earthquake—then I'd find the two situations more analogous. Both have been made worse off by those destructive forces.

It is perhaps telling that at least my intuitions point to examples of possible destruction that most clearly indicate how physical objects, such as paintings and building, can be said to have interests. For when we consider RoN and think about the particular rights that nature may be said to have, and not just speculate in a general abstract manner about nature and rights, we find the right to exist among the most pressing and important. Nature's right to exist—the right to not be destroyed—creates duties to protect nature. Additional rights that are often mentioned include the right to thrive, to regenerate, to evolve, rights that create at least negative duties—that we refrain from certain kinds of harmful and destructive behavior, as well as positive duties, including doing what is necessary to maintain the health of ecosystems, of rivers, of forests, etc.[11]

As we've seen, however, the mere existence of interests in a thing isn't sufficient for granting it rights and personhood. This is clearest in cases where we do want to offer some protection to nature—from excessive tourism, wanton destruction, and pollution—but still don't wish to (or at least haven't) make the additional move to granting rights. That is the status for animals in most jurisdictions, where it is illegal to engage in animal abuse, though not because the animal itself has a right against torture. Likewise, we've seen that when discussing AI, it may be that protections are more easily offered to the hardware that AI systems run on than the AI systems themselves.[12] Of course, there are differences across those examples, but one common theme at least is that granting protection without rights—what I spoke of in the Introduction as the lowest kind of legal visibility—tends to occur when we continue

11 A comprehensive list of the various rights that have been asserted for nature can be found at 'Rights of Nature: What Are They? A Plain Language Legal Guide for Earth Citizens' <https://allard.ubc.ca/sites/default/files/2023-10/RON%20Guide%20%20No%202.pdf>

12 Here, we might avail ourselves of concepts discussed earlier and speak of the "health" of embedded AI system, of whether the AI system located in a robot is maintaining its functional integrity, or of any a host of ways of assessing, at a more holistic level, the fit, interplay, and health of an AI system and the physical materials it is embedded in—whether we are imagining it in a robot or simply the system running on a large server farm.

to think about the objects of protection as just that, objects. Or perhaps more helpfully, objects that are property—whether it be the private property of a person's pet dog, or the property of a business that funded the creation of an AI system, or the property of a community or of the State when we turn to protected wilderness areas.

I've argued that if we wanted to change a given entity's legal status—say, from being object to subject—we will be best severed at the outset by determining if that entity has interests, in at least the minimal sense I've developed. If ecosystems, or dogs, or AI systems can be shown to have interests, we have opened the door at least for the possibility of some of those things as having rights and, thus, of personhood. The Legalist strategy I've developed sets a lower bar for the possibility of granting personhood to some of these more controversial entities. But that, of course, creates a dilemma, as it were: The low bar of relying on this thin notion of "interest" does make it easier to get ecosystems and other parts of nature into the domain of subjects, subjects with rights. The flip side, of course, is that those antecedently convinced that such things cannot and should not be persons will use that as an indictment against relying on such a low bar. We'd therefore expect the Realist who makes this move to insist that personhood tracks features of humans and that only in that way can we rule out the absurdity of building-rights. Having gone through this debate at length by now, there is no need to repeat the various back-and-forth arguments. I will add this, however.

If we are engaging the possibility of extending personhood to non-human elements of nature not just out of a theoretical interest to see how flexible the concept person is but are doing so with an eye to addressing the very real problem of environmental destruction, then not only are we forced to confront the theoretical elements about personhood but also the question of what is effective when it comes to change. My discussions about expressive theories of law have tried to show an additional practical power that emerges were we to give nature its own rights and personhood. And if by parity of reasoning that brings along AI as a legal person in its wake, and there are no sufficiently important dis-analogies, then so be it. For my part, I'm inclined at the moment to think we can satisfy the social-economic-political goals that implicate AI without recourse to its legal personhood. As for nature, as I've said repeatedly, the task of confronting and battling climate change and, more generally, the destruction of our environment will require many different kinds of solutions, in many different places, with many different kinds of actors, laws, theories, and strategies. That Weak Legalism emerges as one among many possible effective solutions I take as confirming that it should remain in the toolbox to be used by jurists, legislators, international bodies, activists, the media, and any and all members of the public who wish to chart a different course and find a way to save us from ourselves.

References

Chopra S and White LF, *A Legal Theory for Autonomous Artificial Agents* (University of Michigan Press 2011).

'The History of the Atom—Theories and Models' <https://www.compoundchem.com/2016/10/13/atomicmodels/>

Raz J, 'Legal Rights' (1984) 4 *Oxford Journal of Legal Studies* 1.

'Rights of Nature: What Are They? A Plain Language Legal Guide for Earth Citizens' <https://allard.ubc.ca/sites/default/files/2023-10/RON%20Guide%20%20No%202.pdf>

Wittgenstein L, *Philosophical Investigations* (Basil Blackwell 1953).

Chapter 8

Conclusion

Nicolaus Copernicus's 1543 *De Revolutionibus Orbium Coelestium* (*On the Revolutions of the Heavenly Spheres*) marked a key moment in what I think is usefully described as the gradual dislodging of humanity from the center of all things. In this case, his heliocentric model was a fairly literal dislodging, as the Earth was dethroned from its place at the center of the universe, with other heavenly bodies left to orbit Earth in perpetuity. Instead, Earth was relegated to being simply another celestial body among many, all of which revolved around the Sun. Copernicus's dislodging was to be followed by others. Charles Lyell's writing on geology put the age of the Earth not in thousands of years, as had been "calculated" based on the Bible, but in several hundred million years. Temporally, the existence of humans was a very small period amongst a vast expanse of time that decidedly did not include us. Lyell's discoveries paved the way for further dislodging, most famously Darwin's theory of natural selection, which not only put human on a biological continuum with other living things but also upset a worldview where the point of the universe was, in a word, to grow humans. Instead, humans are but another species that developed out of contingent processes. Our place at the center of things was once again overturned. Further upheavals came from such figures as Nietzsche who argued that even moral systems have a history and which develop and change through social pressures.[1]

Among the lessons we can draw from this profound period of intellectual revolution and its recognition of contingency and the decentering (i.e. dislodging) of humanity from the center of things is that it is easy to conflate what is familiar with what is necessary. The familiarity and ubiquitousness of capitalism can lead one to think it is the only inevitable economic system, until Marx argued otherwise, viewing it as but one economic possibility among others, one which survives not because it is more natural than others or inevitable but because it is in the interests of some to both perpetuate it

1 See, for instance, Friedrich Nietzsche, *On the Genealogy of Morals* (Robert C Holub ed, Michael A Scarpitti tr, Penguin Classics 2013).

DOI: 10.4324/9781003740186-9

and to suggest it is, in fact, natural. That notions of guilt and duty are inevitable, unavoidable moral sentiments, until Nietzsche suggested otherwise. That tonal music was the only kind possible, until 12-tonal composers, such as Schoenberg, showed otherwise. That traditional methods of painting and sculpture themselves have alternatives, as Impressionists and Expressionists made clear.

I mention these moments and periods of immense and profound change, which certainly deserve more than this superficial gesturing at, in order to highlight, from a different angle, a major theme of this book. Namely, when we turn to law, whether we are focusing on environmental law and policy in particular, or thinking more broadly about rights and personhood, we are in a position to appreciate another kind of possible dislodging. Indeed, one might argue that a legal dislodging of the kind I've been pointing towards is necessary if we want to achieve particular social-economic-political goals—in our case, the strengthening of legal protections of the environment. Previous dislodgings I've mentioned were possible through, among other things, creative twists of imagination. So, too, our ability to think creatively about law's prospects, and the prospects for legal creativity may very well turn on our ability to use our imagination for what is possible at the foundations of law.

The legal dislodging I have in mind has different layers and aspects. A pithy way to summarize the dislodging I'm contemplating might be to say that we are in a position in the 21st century to *leave behind the view which holds that law is of, by, and for the people.* For just as we are not the center of the universe, spatially or temporally, so, too, perhaps law is not just for us.[2]

Legal Dislodgings

What, after all, is the move away from anthropocentrism to biocentrism and ecocentrism if not a dislodging of humans from the center of things, in this case, of what is valuable? The names of those movements wear the dislodging right there on the surface, as we contemplate whether what is valuable centers on the needs and interests of humans or whether, instead, there is a shift

2 An even stronger view, which I have not attempted to defend here or do more than even mention in passing, is the idea that law perhaps is not only *of and by* people. Thinkers such as Berry and Cullinan, not to mention various Indigenous scholars, may be understood as holding that the universe has its own laws, principles, and rules that we would be wise to follow if we are to fully tackle the challenges confronting our environment. I have settled for a much smaller, albeit still controversial, claim, namely that law is not just *for* people but for all of nature, which, in turn, requires law's persons to include the non-human as well. Here, again, I show my indebtedness to Naffine's work, in particular, her book *Law's Meaning of Life: Philosophy, Religion, Darwin and the Legal Person* (Hart Publishing 2009), which asks the central question, "Who is law for?"

towards making the value of the living world, or perhaps even all of nature, the center of our legal orientation. Put slightly differently, just as humanity has been dislodged from the center of universe—spatially, temporally, and biologically—so, too, we can understand calls to reject anthropocentrism as the project of dislodging humanity from the moral or value center of the universe.

As I hope is clear from previous chapters of this book, I have eschewed reliance on moral notions such as intrinsic value wherever possible. Instead, that is, of speaking of the intrinsic or ultimate value of the natural world, however we choose to construe it, and making the case that nature should have legal rights and personhood of its own because nature is the kind of thing that deserves those legal recognitions because, in a word, nature has what it takes to be granted rights and personhood, I've sought to move past having humanity at the center of the legal system by speaking of nature's *interests*, as opposed to its *value*. Further, I've argued that the notion of interests we need does not require the sentience or intrinsic or ultimate value. Corporations were explored as the paradigmatic case of something have interests without being sentient or being intrinsically valuable.[3]

The movement beyond anthropocentrism that this book explores, again, isn't grounded in a new or alternative moral framework or a new kind of ethics, as has been the method explored by many.[4] Instead, one might picture the dislodging of humanity from the center of law not by reference to the value of the rest of nature but by reference to the *interests* that the non-human part of nature have.

This is, in essence, my version of Legalism, where I've defended what I've called Weak Legalism, which holds that there is one constraint on legal personhood, namely that purported legal persons have interests.[5] The position of Strong Legalism, in contrast, holds that there are no constraints at all on being a legal person and that anything in theory could be. What both versions have in common, and what makes them Legalist positions in the first place, is the idea that legal personhood need not track any of traits that have variously been deemed essential to human beings, be it their rationality, or having a soul, or being sentient, or even just having intrinsic or ultimate value.

We should, for clarity's sake, recognize two strands of the dislodging I'm arguing for. They are deeply entwined but distinguishable and related perhaps

3 I've also used this pared-down notion of interests to tentatively explore the possibility of AI systems having such interests and with it the possibility of their being capable of having rights and personhood.

4 A prime example here would be the project of "Deep Ecology," from figures such as Arne Næss, who have argued that the living environment as a whole has intrinsic value, making it deserving of various basic moral and legal rights.

5 Which again was why I tentatively suggested AI systems, at least the ones we presently have, should not be granted personhood, as they simply do not have interests.

by one being a more specific instance of the other. The more general kind of dislodging is the one just mentioned: the rejection of personhood as requiring elements of humanity is a general thesis about the nature of personhood that removes humanity from the center on the question of who law is *for*. If it is only *interests* that matter for legal personhood, and thus for the having of rights, then humans having personhood and rights is but one among countless categories of things that can stand at a legal center.

A second kind of dislodging should be mentioned as well, and it is a theme I've pointed to at various places in previous chapters, where, again, I show my debt to both Naffine's and Grear's analyses. Here is not humanity in general that is to be dislodged but, instead, the patterns and practices that have put at the center of law's respect and protections a particular instantiation of or picture of who the law is actually for—which class of humans, that is. To give an example, UNDHR declares with lofty, universalizing language that all humans are born with dignity and rights that must be respected. Viz:

> Article 1: All human beings are born free and equal in dignity and rights. They are endowed with reason and conscience and should act towards one another in a spirit of brotherhood.
>
> Article 2. Everyone is entitled to all the rights and freedoms set forth in this Declaration, without distinction of any kind, such as race, colour, sex, language, religion, political or other opinion, national or social origin, property, birth or other status. Furthermore, no distinction shall be made on the basis of the political, jurisdictional or international status of the country or territory to which a person belongs, whether it be independent, trust, non-self-governing or under any other limitation of sovereignty.[6]

As an aspirational document, UNDHR presents a vision that unifies and elevates all of humanity, giving every human dignity, respect, and basic rights, without distinctions made based on who or where they are.

On the flip side, as we know, historically, this has never been the case. There has always been a disparity in the actual rights and dignity afforded humans, with distinctions made along lines of class, race, gender, and sexual identity, if not others.[7] The gradual enshrinement of basic rights that explicitly tackles discrimination no doubt counts as progress. But that progress has to be juxtaposed against the reality of the inequality that still exists and how far we are from the noble ideal that all are equal and should be treated

6 UN General Assembly, Universal Declaration of Human Rights, 217 A (III), 10 December 1948 <https://www.refworld.org/legal/resolution/unga/1948/en/11563> accessed 2 March 2026.

7 Or, as Orwell famously put it, "All animals are equal, but some animals are more equal than others." George Orwell, *Animal Farm* (Penguin Books 2021).

as such. Still, the extent that the legal standing and legal protections for historically marginalized groups are gradually improving gives us our second way of understanding the dislodging I've spoken of. Here is the dislodging of wealthy white heterosexual men from the center of the legal universe. Of course, the process itself is not complete, with some continuing to insist that what I'm counting as progress amounts instead to reverse discrimination.[8]

With these two types of legal dislodging in mind—one seeks to remove from the center particular kinds of humans, while the other aims at a wholesale dislodging of humanity itself from the legal center—it is worth asking about the status and challenges facing each. To my mind, the latter project of dislodging humanity from the legal center when it comes to rights and personhood is clearly the more radical project, and one that has the furthest to go. If nothing else, that speaks to the truly extraordinary effort and creativity that will be required if we are to rethink who/what can be the bearer of rights (and duties), particularly in the case of giving rights and personhood to the non-human aspects of nature.

Theoretically, at least, the language of law and human rights law that derives its mandate from such document as the UNDHR is all inclusive. The failure to meet those inspiring and optimistic words becomes evident when we see how the law has tended to actually function: women as property of their husbands, doctrines of "separate but equal" racial treatment in education and beyond, restrictions on who can vote, the plight of refugees, prohibitions on certain kinds of intimate relations, religious intolerances, and the list goes on. Still, we may try to put a positive outlook to this state of affairs by remembering that an ideal has been articulated and that we can point to examples of progress and, among other things, a greater awareness of and courage to confront the various kinds of discrimination that stand in the way. In sum, we might say that, at least in the lofty, abstract language and goals of law, there has been progress in undermining the notion that by "human," we implicitly mean "wealthy, heterosexual white male," though in practice, the dislodgment is incomplete and ongoing.

As for the idea that law is for humans and only humans, arguably, that has been dislodged at the practical level only at the fringes and only slightly more so at the theoretical level. With the exception of corporate persons, the notion of there being legal persons that are not human is still in its infancy, notwithstanding the rapid development that idea has taken in RoN. Noteworthy, I think, is that even the notion of corporate personhood continues to stir unease. I take as evidence for this from the ongoing, continual insistence that there are *natural persons* on the one hand and mere *artificial persons* on the other, with corporate persons being in the latter category. The alleged

8 As is claimed by various right-wing figures who are opposed to diversity, equality, and inclusion initiatives.

"unnaturalness" of the latter speaks to the power still held in the idea that law's persons are first and foremost human beings, and only with charitable extension is there room for another kind—in this case, corporations and a handful of other "odd" cases of ships, idols, and the like. That is to say, the idea that law's persons could just as naturally be non-humans has made far fewer inroads or, to change metaphors, has not been nearly as successful in dislodging the traditional way of theorizing about personhood.

The Legalist project I've articulated and defended here is one step among many to remedy this. While we might wish to use of the language of nature being "other" or to characterize legal developments as moves away from anthropocentrism, or to speak, as I've done here, of dislodging humanity from the center of legal affairs in a way that mirrors other historical moments of dislodging, there is a shared, deep core across these ways of speaking and theorizing. All of them point to a conviction that existing legal structures are unable to meet the challenges confronting the natural world. To rehearse a distinction from the Introduction, I think it is useful to think of there being different ways in which something can be legally visible. When it comes to the environment, nature does have a kind of visibility when it is given various legal protections, such as requiring environmental impact assessments to be done before building or other commercial activity is allowed in a given portion of nature. Or when areas are set aside as protected areas that prohibit any construction or mining or development. To many, this kind of visibility is insufficient, which raises the question of how else to encode nature into the law that might make for stronger protections. I've explored in some detail the cluster of ideas wherein giving nature its own rights, and giving nature its own personhood, would possibly strengthen nature's visibility in a way that can do a better job of protecting the environment.

As RoN is fairly new, there are, as of yet, mixed but mostly just incomplete verdicts on the efficacy of this strategy. And so with that remains the practical question of which methods of protecting the environment are best: ones that seek to strengthen environmental protections with the tools already available, or whether something new and radical would be better. The safest thing perhaps to say about that question is that there may not be a single answer. Different contexts may be best served by different strategies. That includes the nature of the natural area one is seeking protection for, what the local and domestic legal climate looks like, as well as the local politics. Further connected will be questions of whose land is in question, particularly if it is land traditionally used by Indigenous Peoples. Thus, I think it wise to follow Tănăsescu's view, discussed earlier, that RoN is not monolithic in its motivations or strategies.[9]

9 M Tănăsescu, 'The Rights of Nature as Politics' in D Corriga and M Oksanen (eds), *Rights of Nature: A Re-Examination* (Routledge 2021) 69.

Legalism and Realism: One Last Time

Those questions and issues to the side, this book has set out to make several main points, which I briefly highlight now. I've sought throughout these pages to establish two theses. The first can be understood as the conditional claim that *if* nature is to be granted rights and personhood, then all things considered, its rights and personhood should be grounded in Legalist rather than Realist foundations. The second thesis aims to establish the antecedent, as it were. Nature should, in other words, be given personhood and rights of its own. The bulk of establishing the antecedent is through the notion of interests, which I've claimed is a necessary condition of being a rights-holder and, thus, of being a legal person. The notion of interests I've relied on is a fairly weak one, simply requiring of something that it, in essence, be capable of being made better or worse. Given the very low bar this sets for being a potential rights-bearing entity that I've endorsed, under threat of inconsistency, I have had to accept that a whole realm of other things can be rights-bearers and legal persons. This has put me at odds with those whose accounts of interests is paired with notions of sentience or of intrinsic/ultimate worth. Yes, buildings and works of art, along with nature, can become legal persons, while things like colors, numbers, and probably AI cannot.

As important as showing that nature has what it takes to be a legal person and to have rights has been to my overall project, in many ways, it is the establishing of the conditional claim that is even more important. The contribution to RoN I hope to make, in other words, is demonstrating that RoN is better served by completely severing the connection between the legal concept of a person and various conceptions of what it is to be a person in the metaphysical, biological, religious, and moral sense of the term. It is a mistake, I've argued, to assimilate the non-human parts of nature to the human parts. More generally, the legal conception of a person is only contingently connected with our conceptions of humanity. To echo the points earlier about dislodging, we should not mistake the familiarity and history of connecting personhood with humanity for the idea that the two are necessarily related and connected. To do so, among other ills, is to tie law's hands in a way that stands in the way of drawing on and developing law's creative possibilities.

To be sure, a reorientation in how personhood is conceived is not going to magically solve our environmental crises. Nor is it going to be solved just by giving nature rights and personhood, however those are conceived and whatever the theoretical foundations look like. Hopes for radical change in protecting the environment and saving ourselves from ourselves—if that is not too dramatic a way of putting it—is going to require a multipronged strategy that engages lawyers, scientists, politicians, activists, and artists, to name a few. To recast a well-worn phrase, it will take a village to save the village.

References

Naess A, 'Defence of the Deep Ecology Movement' (1984) 6 *Environmental Ethics* 26.
Naffine N, *Law's Meaning of Life: Philosophy, Religion, Darwin, and the Legal Person* (Hart 2009).
Nietzsche F, *On the Genealogy of Morals* (RC Holub ed, MA Scarpitti tr, Penguin Classics 2013).
Orwell G, *Animal Farm* (Penguin Books 2021).
Tănăsescu M, 'Rights of Nature, Legal Personality, and Indigenous Philosophies' (2020) 9(3) *Transnational Environmental Law* 429 <https://doi.org/10.1017/S2047102520000217>
UN General Assembly, Universal Declaration of Human Rights, 217 A (III), 10 December 1948, <https://www.refworld.org/legal/resolution/unga/1948/en/11563>

Index

Note: numbers in *italics* indicate a figure

For Product Safety Concerns and Information please contact our EU representative GPSR@taylorandfrancis.com Taylor & Francis Verlag GmbH, Kaufingerstraße 24, 80331 München, Germany

Batch number: 10739599

Printed by Printforce, the Netherlands